Advances in Intelligent Systems and Computing

Volume 1019

The series "Advances in Intelligent Systems and Computing" contains publications on theory, applications, and design methods of Intelligent Systems and Intelligent Computing. Virtually all disciplines such as engineering, natural sciences, computer and information science, ICT, economics, business, e-commerce, environment, healthcare, life science are covered. The list of topics spans all the areas of modern intelligent systems and computing such as: computational intelligence, soft computing including neural networks, fuzzy systems, evolutionary computing and the fusion of these paradigms, social intelligence, ambient intelligence, computational neuroscience, artificial life, virtual worlds and society, cognitive science and systems, Perception and Vision, DNA and immune based systems, self-organizing and adaptive systems, e-Learning and teaching, human-centered and human-centric computing, recommender systems, intelligent control, robotics and mechatronics including human-machine teaming, knowledge-based paradigms, learning paradigms, machine ethics, intelligent data analysis, knowledge management, intelligent agents, intelligent decision making and support, intelligent network security, trust management, interactive entertainment, Web intelligence and multimedia.

The publications within "Advances in Intelligent Systems and Computing" are primarily proceedings of important conferences, symposia and congresses. They cover significant recent developments in the field, both of a foundational and applicable character. An important characteristic feature of the series is the short publication time and world-wide distribution. This permits a rapid and broad dissemination of research results.

** Indexing: The books of this series are submitted to ISI Proceedings, EI-Compendex, DBLP, SCOPUS, Google Scholar and Springerlink **

More information about this series at http://www.springer.com/series/11156

Alexander Palagin · Anatoliy Anisimov ·
Anatoliy Morozov · Serhiy Shkarlet
Editors

Mathematical Modeling and Simulation of Systems

Selected Papers of 14th International
Scientific-Practical Conference, MODS, 2019
June 24–26, Chernihiv, Ukraine

 Springer

Editors
Alexander Palagin
VM Glushkov Institute of Cybernetics
Ukraine National Academy of Science
Kyiv, Ukraine

Anatoliy Anisimov
Taras Shevchenko National
University of Kyiv
Kyiv, Ukraine

Anatoliy Morozov
Institute of Mathematical Machines
and Systems Problems
Ukraine National Academy of Science
Kyiv, Ukraine

Serhiy Shkarlet
Chernihiv National
University of Technology
Chernihiv, Ukraine

ISSN 2194-5357   ISSN 2194-5365   (electronic)
Advances in Intelligent Systems and Computing
ISBN 978-3-030-25740-8   ISBN 978-3-030-25741-5   (eBook)
https://doi.org/10.1007/978-3-030-25741-5

This Springer imprint is published by the registered company Springer Nature Switzerland AG
The registered company address is: Gewerbestrasse 11, 6330 Cham, Switzerland

Preface

The International Scientific-Practical Conference "Mathematical Modeling and Simulation of Systems" (MODS) was formed to bring together outstanding researchers and practitioners in the field of mathematical modeling and simulation from all over the world to share their experience and expertise.

It was established by Institute of Mathematical Machines and Systems Problems of the National Academy of Sciences of Ukraine (IMMSP of the NASU) in 2006. MODS is an annual international conference organized by IMMSP of the NASU and Chernihiv National University of Technology (CNUT). Since 2018, the State Scientific Research Institute of Armament and Military Equipment Testing and Certification became one of the conference organizers.

The XIVth International Scientific-Practical Conference MODS'2019 was held in Chernihiv, Ukraine, during June 24–26, 2019.

During this conference, technical exchanges between the research community were carried out in the forms of keynote speeches, panel discussions, as well as special session. In addition, participants were treated to a series of receptions, which forge collaborations among fellow researchers.

MODS'2019 received 64 papers submissions from different countries. All papers went through a rigorous peer-review procedure including pre-review and formal review. Based on the review reports, the program committee finally selected 25 high-quality papers for presentation at MODS'2019, and which are included in "Advances in Intelligent Systems and Computing" series.

This book contains papers devoted to relevant topics including tools and methods of mathematical modeling and simulation in ecology and geographic information systems, manufacturing and project management, information technology, modeling, analysis, and tools of safety in distributed information systems, mathematical modeling and simulation of special purpose equipment samples. All of these offer us plenty of valuable information and would be of great benefit to the experience exchange among scientists in modeling and simulation.

The organizers of MODS'2019 made great efforts to ensure the success of this conference. We hereby would like to thank all the members of MODS'2019 Advisory Committee for their guidance and advice, the members of program committee and organizing committee, and the referees for their effort in reviewing and soliciting the papers, and all authors for their contribution to the formation of a common intellectual environment for solving relevant scientific problems.

Also, we grateful to Springer-Verlag and Janusz Kacprzyk as the editor responsible for the series "Advances in Intelligent System and Computing" for their great support in publishing these selected papers.

Alexander Palagin
Anatoliy Anisimov
Anatoliy Morozov
Serhiy Shkarlet

Organization

Organizers

Ministry of Education and Science of Ukraine
The National Academy of Sciences of Ukraine
Academy of Technological Sciences of Ukraine
Engineering Academy of Ukraine
State Scientific Research Institute of Armament and Military Equipment Testing and Certification, Ukraine
Glyndwr University, Wrexham, UK
US Army Research Laboratory, USA
Defence Institute of Tsvetan Lazarov, Bulgaria
Lodz University of Technology, Poland
Riga Technical University, Latvia
Tallinn University of Technology, Estonia
University of Extremadura, Badajoz, Spain
Francisk Skorina Gomel State University, Belarus
Institute of Mathematical Machines and Systems Problems of the NASU
Keldysh Institute of Applied Mathematics of the RAS, Moscow, Russia
National Technical University of Ukraine "Kyiv Polytechnic Institute"
Yuriy Kondratyuk Poltava National Technical University, Ukraine
The Bohdan Khmelnytsky National University of Cherkasy, Ukraine
Chernihiv National University of Technology, Ukraine

Chairs

Ireneusz Zbiciński	Lodz University of Technology, Poland
Morozov A. A.	Institute of Mathematical Machines and Systems Problems of the NASU, Ukraine
Onishchenko V. A.	Yuriy Kondratyuk Poltava National Technical University, Ukraine

Enrique Romero-Cadaval	University of Extremadura, Badajoz, Spain
Shkarlet S. N.	Chernihiv National University of Technology, Ukraine
Bashynskyi V. G.	State Scientific Research Institute of Armament and Military Equipment Testing and Certification, Ukraine
Vasiliev A. I.	A. N. Podgorny Institute for Mechanical Engineering Problems of the NASU, Ukraine
Dmitri Vinnikov	Tallinn University of Technology, Estonia
Ilya Galkin	Riga Technical University, Latvia
Demidenko O. M.	Francisk Skorina Gomel State University, Belarus
John Davies	Glyndwr University, Wrexham, UK

Program Committee

Adamchuk V. V.	The National Academy of Agrarian Sciences of Ukraine, Ukraine
Azarov O. D.	Vinnytsia National Technical University, Ukraine
Verlan A. F.	National Technical University of Ukraine "Igor Sikorsky Kiev Polytechnic Institute," Ukraine
Voloshyn O. F.	Taras Shevchenko National University of Kyiv, Ukraine
Gitis V. B.	Donbass State Engineering Academy, Ukraine
Holub S. V.	Yuriy Fedkovych Chernivtsi National University, Ukraine
Golubev Y. F.	Lomonosov Moskow State University, Russia
Gryshko V. V.	Yuriy Kondratyuk Poltava National Technical University, Ukraine
Dmytriiev V. A.	State Scientific Research Institute of Armament and Military Equipment Testing and Certification, Ukraine
Zhelezniak M. Y.	Institute of Mathematical Machines and Systems Problems of the NASU, Ukraine
Janis Zakis	Riga Technical University, Latvia
Kazymyr V. V.	Chernihiv National University of Technology, Ukraine
Kharchenko V. S.	National Aerospace University named after N.E. Zhukovsky "Kharkiv Aviation Institute"
Klimenko V. P.	Institute of Mathematical Machines and Systems Problems of the NASU, Ukraine
Kovalevskyi S. V.	Donbass State Engineering Academy, Ukraine

Kovalets I. V.	Institute of Mathematical Machines and Systems Problems of the NASU, Ukraine
Kostogryzov A. I.	Gubkin Russian State University of Oil and Gas, Russia
Kraskevych V. E.	Kyiv National University of Trade and Economics
Lytvynov V. V.	Chernihiv National University of Technology, Ukraine
Liakhov O. L.	Yuriy Kondratyuk Poltava National Technical University
Maderych V. S.	Institute of Mathematical Machines and Systems Problems of the NASU, Ukraine
Myronenko V. G.	The National Academy of Agrarian Sciences of Ukraine, Ukraine
Mozharovskiy V. V.	Francisk Skorina Gomel State University, Belarus
Sagayda P. I.	Donbass State Engineering Academy
Snytiuk V. E.	Taras Shevchenko National University of Kyiv, Ukraine
Stetsenko I. V.	National Technical University of Ukraine "Igor Sikorsky Kiev Polytechnic Institute," Ukraine
Tarasenko V. P.	National Technical University of Ukraine "Igor Sikorsky Kiev Polytechnic Institute," Ukraine
Tarasov O. F.	Donbass State Engineering Academy (DSEA)
Tomashevskiy V. M.	National Technical University of Ukraine "Igor Sikorsky Kiev Polytechnic Institute," Ukraine

Steering Chairs

Igor Skiter	Institute of Mathematical Machines and Systems Problems of the NASU, Ukraine
Igor Brovchenko	Institute of Mathematical Machines and Systems Problems of the NASU, Ukraine

Local Organizing Committee

Alla Hrebinnyk	Institute of Mathematical Machines and Systems Problems of the NASU, Ukraine
Mariia Voitsekhovska	Chernihiv National University of Technology, Ukraine

Sponsors

Ministry of Education and
Science of Ukraine

«S&T Ukraine», Kyiv, Ukraine

Cyber Rapid Analysis for
Defense Awareness of Real-
time Situation – CyRADARS
(Project SPS G5286)

Contents

Mathematical Modeling and Simulation of Systems in Ecology and Geographic Information Systems

Geoinformation Technology of Data Processing in the Forestry Industry on the Example of a System "GIS-Lisproekt"

Svitlana Ya. Maistrenko$^{(\boxtimes)}$ ⓘ, Taras A. Doncov-Zagreba ⓘ,
Viacheslav P. Bespalov ⓘ, and Kostyantyn V. Khurtsilava ⓘ

Institute of Mathematical Machines and Systems Problems,
NASU, Kyiv, Ukraine
maistrsv@ukr.net

Abstract. The geoinformation technology of processing of cartographic and inventory information of forest management is described in the article, using the example of the "GIS-Lisproekt" system developed by the authors. Particular attention is paid to the post-processing of primary geospatial and inventory data of the forests and its accumulation in the geobase of the server type. The structure of the geobase has been designed while taking into account the hierarchical subordination of the subjects of forest management. As a base, a universal GIS - ArcGis from ESRI is chosen. In order to expand the functionality of a general-purpose GIS system, specialized tools for solving planning and cartographic support issues for forest sector management are proposed and developed. Arguments in support of the need to develop specialized tools and features of their implementation are given. A specialized component that allows simultaneous updates of "logically" related layers, as well as an automatic change of inventory and other semantic characteristics of forest allotments, are proposed. In conclusion, attention is paid to the need to ensure public access to the geospatial information resources related to forests and the possibility of providing it through the use of ArcGis as a basic GIS, data storage in a server-type geobase and their actual accumulation over various time periods. This, in turn, makes it possible to perform geospatial analysis and use its results in decision support systems for the forest sector.

Keywords: GIS technology · Geobase · Geoportal · Forest management

1 Introduction

In contemporary conditions, geographic information has become an important strategic resource of public administration, a significant factor in the socio-economic development of the state and its integration into the global information space.

Most countries of the world have developed and implemented programs for the creation of the national geospatial data infrastructures. Such programs are aimed at improving the system of meeting the needs of society in all types of geographic information, increasing the efficiency of using geospatial data and technologies in

© Springer Nature Switzerland AG 2020
A. Palagin et al. (Eds.): MODS 2019, AISC 1019, pp. 3–12, 2020.
https://doi.org/10.1007/978-3-030-25741-5_1

decision support systems of the state authorities (or local governments) in the economic, social, environmental, defense, scientific fields, etc. [1].

Modern geoinformation systems (GIS) are focused on providing support for making optimal management decisions based on spatial data analysis. This kind of data makes up more than half the volume of all information resources in organizations that are engaged in various activities that require consideration of the spatial location of objects. One of the areas of a wide application of geoinformation technologies is the country's forest management, where geosystems provide the possibility of effective management, monitoring of forest resources and control of use, restoration and turnover of the forest fund.

While creating a GIS for the forest sector, various universal GIS of well-known manufacturers from the USA, Canada, Russia, are used as the base ones, such as ArcGis, MapInfo, Panorama, and others. But at the same time, there is also "pure" - specialized GIS, which is designed for the "unique" needs of automating forest management processes. These are systems such as Topol - L [2], LUGIS [3], LesGIS [3], GIS "Forest Resources" [4] and others. Specialized systems allow obtaining integrated cartographic and forest allotments information for solving practical problems of forest inventory, current planning of forest cutting, reforestation, fire and forest protection measures, creating various thematic forest maps, making current changes in the forest fund, etc.

This article is devoted to the description of the developed and implemented geoinformation technology for obtaining, accumulating, storing, updating and providing cartographic and attributes inventory information for forest management, using modern technology and software tools and technologies using the example of the "GIS-Lisproekt" system developed by the authors of the article. During development of this geoinformation technology and the "GIS-Lisproekt" system, certain technological aspects and individual components of the system were covered in previous publications [5]. This article for the first time provides a comprehensive description of geoinformation technology and technical solutions to problematic issues that have arisen in the process of creating a system for the State Enterprise "Ukrderzhlisproekt".

2   Accumulation, Processing, and Storage of Forest Inventory Data

To control and monitor a variety of anthropogenic and natural objects, as a rule, remote sensing technologies (Earth remote sensing) are used, utilizing ground-based, aviation and space tools equipped with various types of imaging equipment. Forestry is one of the areas of application of such technologies for obtaining primary reliable information on the state of forests, which allows solving economic and environmental problems.

Figure 1 presents an example of a scheme containing a set of subsystems that allow for solving forest management problems as it is done for example by the State Enterprise "Ukrderzhlisproekt". This enterprise carries out forest management in a unified way throughout Ukraine.

Blocks 1–3 allow solving the tasks of obtaining primary cartographic and forest inventory data, as well as accumulation of inventory and other attribute information. Blocks 4–5 are intended for specialized post-processing of cartographic information for the direct solution of forest management problems, as well as the possible provision of public access to the accumulated data.

Block 1 is used to receive input data, using remote sensing (aerial photograph of the terrain with the subsequent processing of images at a digital photogrammetric station [6]). In total, as a result of digitization, 255 primary layers are provided.

Block 2 allows you to receive materials of field surveys of the territories (data of topographical, engineering and geodetic studies, etc.).

Block 3 is used to accumulate and store attribute data (inventory descriptions, forest inventory, etc.).

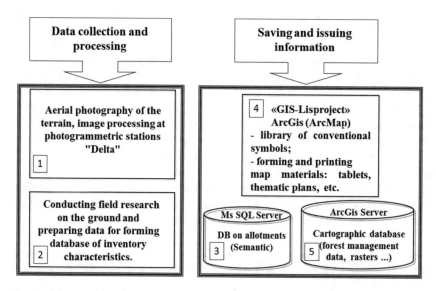

Fig. 1. Scheme of functional modules of the production association "Ukrderzhlisproekt".

Block 4 is intended for post-processing of layers obtained as a result of processing aerial photography. Primary layers must be grouped together. As a result, up to 103 aggregated layers can be obtained, where uniform thematic layers are combined for ease of subsequent use. For example, gas pipelines, power lines, communication lines, oil pipelines, and water mains are combined into one thematic layer - communications. Also, this unit is responsible for automating the process of creating a package of map-materials (tablets of forestry subdivisions, thematic plans of forestry subdivision and thematic maps of forestry division and region of the country).

Block 5 is a component that provides for creation and maintenance of a carto-graphic database of the server type, with the possibility of automated query generation.

For the post-processing of inventory and primary cartographic information, the authors of the article developed a "GIS-Lisproekt" geoinformation system. As a base, a modern full-featured GIS - ArcGis from ESRI was used, supplemented with specialized functional modules that extend the standard capabilities with the necessary functions for solving forestry problems.

The system "GIS-Lisproekt" has developed a structure of an object-relational geo-database of server type [7], which is based on the hierarchical subordination of forest management entities and consists of forestry data tables, geometric data tables and service information tables for the formation of map-materials. Hierarchical code of forestry subdivision includes codes: department, region, regional management, forestry division, subdivisions.

Data tables are a kind of reference books, such as a list of regions, regional administrations, forest inventory management, forestry divisions, forestry subdivisions, forest quarters, forest allotments.

Tables of geometric data, depending on the characteristics of the object, are tied accordingly to the forest allotment, forest quarters or forestry subdivision. Thus, in these tables, each record, depending on the type of geometry, characterizes the forest allotments, forest quarters, or forestry subdivision, for which, in turn, the year of forest management, the forestry division, the regional management, and the region of the country are defined. Geometric tables are created for such objects as polygonal forest allotments, notional glades, posts, rivers, fire breaks, roads, etc.

This geobase structure allows you to organize query generation by using ADO (ActiveX Data Objects) technology. As a result, we obtain classes of the geometric description of the terrain (FeatureClass), which can be assigned to thematic layers, add these layers to maps and save in shapefile format. In the "GIS-Lisproekt" system, a mechanism for automated query generation has been developed and implemented. The user can get: data for all forestry subdivision or some of its layers; data on the part of forestry subdivision (specified forest quarters, forest allotments) (See Fig. 2); "Glued" data for the entire forestry division or the entire region by thematic layers.

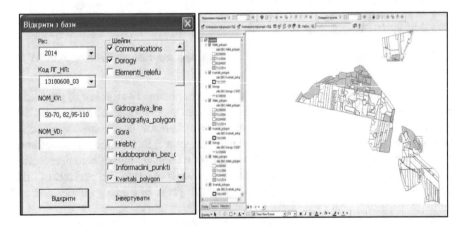

Fig. 2. Example of the selection of individual layers and separate objects of a layer.

Creating a specialized and at the same time simple user interface in the GIS environment greatly simplifies the process of building queries and enhances the functionality of the system. The feasibility of using MS SQL Server, ArcSDE (Spatial Database Engine) and ArcGis Server software from Esri while creating a GIS database of forest management is determined by their functionality. So, due to the multi-user client/server architecture, ArcSDE allows you to instantly perform complex spatial queries, and the total number of clients simultaneously requesting queries has virtually no effect on performance. In addition, ArcSDE uses a compressed binary format for storing geographic data and can work with spatial modules of some DBMSs designed to store and manage the geometric characteristics of objects. In these cases, the geometry also becomes directly accessible through the corresponding SQL implementation for a specific DBMS.

3 The Use of Standard Features of Full-Featured GIS and the Need to Develop Specialized Tools for Solving Problems of Forest Management

As it is known, the technological structure of automated cadastral systems, providing information transformations of cadastral information, consists of three main modules: data collection and processing, storing and issuing cadastral information, modeling and issuing cadastral information. The "GIS-Lisproekt" system can be notionally considered as a prototype of the cadastral system. In addition to the accumulation and issuance of cartographic and associated inventory information at the request of the user, the system provides for the formation of map-materials required in the management of forestry.

Modern standards of production and replication of planning and cartographic materials of forest management in Ukraine and other countries of the former Soviet Union determine the list of requirements for the formation of the main types of maps-materials: tablets, thematic plans of the forestry subdivisions, and schematic maps of the forestry divisions and regions. These are such requirements as application of symbols in accordance with the classifier and technological instruction; printing map-materials by type of atlas; division of the map into separate sheets (the position and number of sheets is determined by the user interactively); ensuring the possibility of arranging combined sheets from geographically distant parts of thematic layers while preserving the coordinate reference; definition of the image border on a separate sheet by the contour (limit) of the forest quarters located in it; formation and output of labels in the form of a fraction: $\frac{Indicator\,1}{Indicator\,2}$" or "$\frac{Indicator\,1}{Indicator\,2:Indicator\,3}$", line: "$Indicator1$" or "$Indicator1 - Indicator2$"; ability to change for individual labels of the same thematic layer of the type of output and point of placement, etc.

The study of standard instruments of ArcGis, as well as analysis of the capabilities of the above mentioned specialized GISs [2–4], showed that their use without additional functionality will not ensure meeting these requirements. Therefore, in the "GIS–Lisproekt" system, specialized modules have been developed to accommodate particularities of the national forest inventory management.

In order to correctly display the thematic layers of the map materials in the "GIS-Lisproekt" system, a Classifier of symbols of forest inventory objects has been created, compatible with the state classifier of topographic information of Ukraine, and software tools for using classification in ArcGis environment. In particular, a mechanism has been developed for combining four or more classifier characters. For example, the combination of such signs as, "contour of forest allotments", "plantings on damp and wet places", "main breed", a sign "not closed forest cultures" and presence of "log cabins" for one forest allotment.

For creating atlases or albums of maps, ArcGis provides a toolkit for multi-page layout (Data Driven Pages). However, this toolkit does not meet the necessary conditions for placement of forestry quarters on the sheets obtained during the "cutting". Namely, for tablets: the forest quarter should "fall" on the whole sheet; for plans: a forest quarter can "fall" on the sheet either wholly or partially. Besides, there is no possibility in that to create combined sheets with preservation of the coordinate reference of objects. In "GIS-Lisproekt", a specialized cutting mechanism is implemented, providing additional features.

The standard features of ArcGis also do not allow (while preparing individual sheets for printing) the removal of "extra", for a specific sheet of map, fragments and do not provide storage of extra-frame design sheets (tablet number, region, forestry division, forestry subdivision, scale, party leader, surveyor, engineer, etc.). These functions are realized in the "GIS-Lisproekt" system. Figure 3 shows an example of "cutting out" forestry subdivision into sheets for the formation of tablets and the layout of one of the tablets prepared for printing.

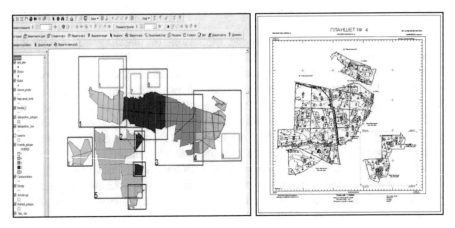

Fig. 3. Example of the division of forestry subdivisions into tablets (left) and the layout of a separate tablet (right).

There is a basic possibility for specifying the same type of format and placement of all the labels for each thematic layer (Layer Properties) for displaying inscriptions on maps-materials in ArcGis, which is not always sufficient. In the "GIS-Lisproekt" system, it is possible to select different types of labels for one thematic layer and change the placement points for individual objects of the layer at the user's choice [6]. For example, in Fig. 4, for the 9 forest allotment of 83 forest quarter, the label is specified as a fraction (number/area), for forest allotment 12 - a dash (number - area), and for forest allotment 20, only the number is defined.

 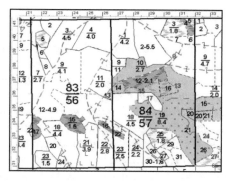

Fig. 4. Selection of type of labels for various objects of one thematic layer; the menu allows for the following choices of the type of signature: 'fraction'; 'using dash', 'string', 'note (fraction)', 'note (using dash)', 'note (string)'.

The system provides for the formation of such basic types of maps-materials: tablets (scales 1:10,000, 1:5,000); thematic plans of forestry subdivisions (scales 1:25000, 1:10000); thematic maps for forestry divisions and regions of Ukraine (scales 1:100000, 1:200000). Thematic forest maps are formed to display the distribution of plantings by species and age groups, forest types, soil types, types of soil-forming underlying subsoils, economic, fire prevention measures and other indicators. Figure 5 shows a fragment of a thematic plan of a forestry subdivision by forest type and the corresponding legend.

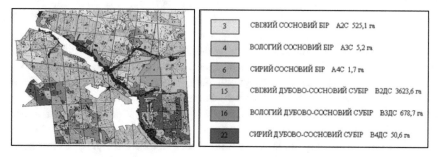

Fig. 5. Fragment of a thematic plan and legend by forest types; the following types are described by the legend: 'fresh pine forest', 'wet pine forest', 'damp pine forest', 'fresh oak-pine forest', 'wet oak-pine forest', 'damp oak-pine forest'.

4   Tasks and Solutions for Updating Cartographic Information to Ensure the Continuity of the Forest Inventory Process

In order to ensure the actualization of Ukraine's forestry inventory, in the production association "Ukrderzhliproekt", data is stored at regular intervals of one year. During this period it is possible to change both the inventory and cartographic characteristics of the objects of thematic layers of a forestry subdivision, as units of information storage.

The system "GIS-Lisproekt" has implemented the function of making changes to the cartographic database. In traditional technologies [6] the source for making changes to the cartographic database is: the database of vector data of forestry inventory of the past years; previous wood inventory raster tablets with manual changes (design of the forest allocations, changes resulting from economic activity or natural phenomena, etc.); records of cartographic changes in the borders of forestry allotments for the specified period, etc. For the same purpose, aerial photographs (for example, raster images), etc., are also used. ArcGis (ArcView/ArcEditor) standard tools allow you to perform various actions to change the geometric characteristics of objects in the individual thematic layers of the map. However, in that, there is no possibility of simultaneously changing the "logically" connected layers. In practice, the object of operations for changing geometry can be both a forestry subdivision as a whole, as well as its separate components. These can be forest quarters, polygonal forest allotments (lakes, rivers, highways …) and linear forest allotments (roads, linear rivers …), etc. For example, when a user performs a division of a forest quarter or several neighboring forest quarters, it is necessary to simultaneously perform an automatic or semi-automatic division of forest quarter related thematic layers (allotments, roads, lakes, communications, etc.).

The "GIS-Lisproekt" system supplements the standard features of ArcGis with such functions as: synchronization of changes to the inventory and cartographic database; ensuring the removal of the selected area of a forestry subdivision; the transfer of the selected area of a forestry subdivision for incorporation into another forestry subdivision; replacement of numbers of forest quarters, forest allotments; replacement of the type of forestry subdivision border (instrumental or visual); division of the polygonal elements of forestry subdivision, including those consisting of several polygons of the cut layer (one object of a layer consisting of several polygons). Usually, this is used to transform "composite" forest quarters and "composite" tracts of a forestry subdivision.

Thus, in the "GIS-Lisproekt" system, with the help of the ArcObjects libraries, tools adapted for editing the data of the forest inventory objects are implemented. It should be noted that this system has designed and implemented a mechanism for making changes not only in the cartographic database but also in the database of maps-materials of a forestry subdivision, created in the previous period of time. Namely, it provides an automatic change of inventory and other semantic characteristics of the forestry subdivisions based on changes in the inventory database; automated changes in the division of the forestry subdivisions into sheets; automated change of information for the formation of labels of maps-materials, etc.

5   Providing Public Access to Geospatial Information Resources

In the infrastructure of geospatial data of developed countries, geoportals are considered as the main means of exchanging geodata between state geographic information systems and the provision of various government services based on electronic government (e-government) on the Internet.

In Ukraine, taking into account the growth of available geodata of various cadastres, as well as the development of e-management systems with functions to provide information services in the field of geo-portal building and formation of the national geospatial data infrastructures, there is a steady trend in the need for modern geo-portal development tools.

Among the created and developed geoportals, there are five main groups [8]: general reference geoportals with metadata catalogs of geoinformation resources of the infrastructure of the corresponding level; cadastral geoportals providing access to public data of certain species inventories with elements of e-government; specialized sectorial geoportals by types of geodata profile sets and specialized services for their online use (environmental, tourism, etc.); geoportals of access to digital satellite images and other materials of remote sensing of the Earth; complex international geoportals for monitoring individual projects.

A geoportal organization provides for three main components: a metadata catalog on the portal, where users search for data and post information about available data; GIS nodes where users place spatial data; GIS users who directly search and post data. According to the basic concept of portal organization of spatial data, the basic principles of creating geoportals are uniform data creation and their most effective support, available possibility of combining spatial data from different sources, ease of data transfer, accessibility, easy search, assignment assessment for specific purposes, and suchlike. Geoportal as a complex system requires the availability of various types of software, according to the computer-aided design system: legal, methodical, software, geo-service, linguistic, informational, etc.

The problematic issues of creating geoportals are the choice of software since the majority of off-the-shelf solutions requires significant funds for the purchase and maintenance of software; availability of qualified specialists to work with specialized programs; electronic maps maintenance in an up-to-date state.

The "GIS-Lisproekt" system is implemented as a set of additionally developed components to ESRI USA standard ArcGis tools, and the use of a server-type geo-database for storing information at different time points provides the ability to create a geoportal using ArcGIS Geoportal Server [9].

The functioning of a geoportal can be based on the use of a metadata database collected in thematic catalogs in accordance with the hierarchical structure of the subjects of forest management. This geoportal should provide search and viewing of information resources and WEB-services by metadata; access to information resources and download of copies using the international standard Web Feature Service: ISO 19142 WFS, as well as other open specifications, in particular, KML; transformation of spatial data, including data coordinate system transformation services. Application of this approach will ensure full compatibility of the system with international standards and specifications of the Open GIS Consortium, of which ESRI is an active participant.

6   Findings

The described technology includes the entire chain of processing cartographic and inventory information of the forest inventory: receiving, primary processing, post-processing, accumulation, updating and issuing on demand in the form of thematic layers, individual layer objects, maps-materials, reports, etc.

The effectiveness of the described technology is confirmed by the successful long-term use of the "GIS-Lisproekt" system software by the production association "Ukrderzhlisproekt" for forest management throughout Ukraine.

Creating a system as a specialized add-on to a fully functional scalable geoinformation platform ArcGis allows full use of both standard and specialized GIS functions.

The use of a geodatabase of the server type for storing cartographic characteristics of the forestry subdivisions and the developed subsystem of access to information provides the ability to perform spatial analysis for solving forest management tasks based on actual data. It also simplifies the task of creating a data geoportal for providing public access to geospatial information resources.

The proposed technology can be used in the development of monitoring systems for the environment, forest fires, prediction of atmospheric pollution as a result of technogenic accidents [10] and other systems that require processing large amounts of spatial information and ensuring that relevant specialists and the public have access to it via the Internet.

References

1. Cherin, A.G., Gorkovchu, M.V.: Structure and functions of the geoportal of the state geodesic network of Ukraine. News Geodesy Cartogr. 1, 24–27 (2013)
2. Geographic Information System (GIS) TopoL. http://www.lesis.ru
3. Martynov, A.N., Melnikov, E.S., Kovyazin, V.F.: Fundamentals of forestry and forest inventory. In: Martynov, A.N. (ed.) Publishing House "Lan", St. Petersburg (2008)
4. GIS "Forest Resources". http://www.belinvestles.by/GIS.html
5. Biletsky, B.O., A., V., Bespalov, V.P., Zagreba, T.A., Maistrenko, S.Y.: "GIS - Lisproekt" - the main technological features, stages of development and development prospects. In: Materials of the Scientific-Practical Conference with International Participation "Decision Support Systems. Theory and Practice, Kiev, pp. 144–147, 8 June 2012
6. Digital photogrammetric station "Delta". http://www.vingeo.com/Rus/delta.html/
7. Tikunov, V.S., Kapralov, E.G., Koshkarev, A.V.: Fundamentals of geoinformatics. In: Tikunov, V.S. (ed.) Publishing Center "Academy", Moscow (2008)
8. Karpinsky, Y.O., Lyashchenko, A.A., Cherin, A.G.: Network of geoportals of the national infrastructure of geospatial data. Cartography 5, 70–74 (2012)
9. Geoportal Server. https://www.esri.com/en-us/arcgis/products/geoportal-server/overview
10. Kovalets, I.V., Maistrenko, S.Y., Khalchenkov, A.V.: Web-based software system 'Povitrya' of operational atmospheric pollution forecasting in Ukraine following technogenic accidents. Sci. Innov. 13, 11–22 (2017)

Classification of Radioactivity Levels in the Regions of the World Ocean Using Compartment Modelling

Vladimir Maderich[1], Roman Bezhenar[1(✉)], Igor Brovchenko[1],
and Vazira Martazinova[2]

[1] Institute of Mathematical Machine and System Problems,
Glushkova av., 42, Kyiv 03187, Ukraine
vladmad@gmail.com, romanbezhenar@gmail.com
[2] Ukrainian Hydrometeorological Institute, Nauki av., 37, Kyiv 03028, Ukraine

Abstract. The time dependent concentrations of ^{137}Cs calculated by compartment model POSEIDON-R for two ocean marine basins (Black Sea and northeastern Atlantic shelf) for period 1945–2020 were classified using the "etalon" method to provide representative distribution of the radionuclide monitoring data for these areas. Application of the classification method allowed to distinguish 4 regions for Black Sea and 11 regions for northeastern Atlantic shelf. The most representative for allocated regions "etalon" boxes were also determined. The developed approach can be used for selection of areas for long-term monitoring of radioactivity levels as most representative locations.

Keywords: Marine radioactivity · Classification · Compartment model

1 Introduction

The distribution of anthropogenic radionuclides in the marine environment varies due to different sources of pollution (global fallout in the atmosphere and regional contamination of ocean caused by the nuclear weapon tests, releases from nuclear reprocessing plants, Chernobyl and Fukushima accidents), and due to differences in marine basins shape and ocean circulation. The data of monitoring of the radioactivity concentration in the World Ocean were collected in frame of several projects of International Atomic Energy Agency (IAEA): MARDOS [1], WOMARS [2] and ongoing LAMER (2017–2021).

In frame of the MARDOS project [1] the radioactivity levels of ^{137}Cs in the sea water have been estimated for the FAO (Food and Agriculture Organization) fishing areas. For areas in which ^{137}Cs distributions were reasonably homogenous, a mean of the existing data set was determined. However, two regions (the northeast Atlantic, and the Mediterranean and the Black Seas) were essentially non-homogeneous in terms of their ^{137}Cs distributions. These regions were divided into sub-regions (Baltic Sea, Danish Straits, North Sea, Faroese Waters, Norwegian/Greenland, Iceland, and Irish Sea), for which a representative ^{137}Cs activity was determined and then a representative ^{137}Cs activity was chosen for each entire region.

© Springer Nature Switzerland AG 2020
A. Palagin et al. (Eds.): MODS 2019, AISC 1019, pp. 13–20, 2020.
https://doi.org/10.1007/978-3-030-25741-5_2

The observation data in the project WOMARS [2] were grouped in the different latitudinal boxes chosen on the basis of known ocean current systems, the location of nuclear test sites and nuclear reprocessing plants, the availability of recent data and, the probability of a relatively uniform distribution of radionuclides. In the Atlantic ocean a several local boxes were added to take into account regional sources: reprocessing plants (Sellafield and La Hague) and Chernobyl accident fallout. However, as noted in [2]: "It is difficult to give meaningful average concentrations for these radionuclides in marginal seas such as the North Sea, the Baltic Sea, the Black Sea and especially, the Irish Sea". The Pacific and Indian oceans also have been divided into different latitudinal boxes. Twelve boxes have been delineated in the Pacific Ocean, three has been allocated for the Indian Ocean and one box has been attributed to the Southern Ocean. The Sea of Japan/East Sea has been given a separate box because of its oceanographic specificity and importance in marine radioactivity studies. However, this regional division did not take into account test sites in the Pacific and Indian oceans and essentially inhomogeneous and time-varying contamination due to the Fukushima accident.

Therefore, more detailed division of marine basins is necessary to provide representative distribution of radionuclides for these areas. In this study, the time dependent concentrations of ^{137}Cs calculated by compartment model POSEIDON-R [3, 4] for two ocean marine basins (Black Sea and North-Eastern Atlantic shelf) where regular releases and/or large scale accidents took place were classified using the "etalon" method [5].

2   Model

The POSEIDON-R [3, 4] is a box model where the marine environment is modelled as a system of compartments for the water column, bottom sediment and biota. The POSEIDON-R 3D compartment model simulates transfer of radioactivity in the water column and bottom sediment. The water column compartment is vertically subdivided into layers. The suspended matter is settling in the water column. The radionuclide concentration in the water compartment is governed by a set of differential equations that describe the temporal variation in the concentration, the exchange of radionuclides between adjacent compartments and between radionuclides in suspension and in the bottom sediment, and radioactivity sources and decay. Exchanges among the water column layers are described by radionuclide fluxes due to advection, sediment settling, and turbulent diffusion. The bottom sediments are divided into three layers, and the transfer of radioactivity between the upper sediment layer and the water column resulting from resuspension, diffusion and bioturbation, and between the upper and middle sediment layers, resulting from diffusion only, are described. Downward burial processes operate in all three sediment layers.

3   Classification Method

The purpose of classification of time series of radionuclide concentration in the set of the boxes is to select groups of boxes with similar temporal distribution of concentration. The selection method is based on "etalon" approach [5]. This selection is to be accomplished in the sequence of steps (see Fig. 1).

Step 1. The data are represented as a matrix with elements x_{ij}, where $1 \leq i \leq n$, $1 \leq j \leq m$; n is number of boxes; m is number of terms in time series.

Step 2. The time mean for each box is calculated as

$$\bar{x}_i = \frac{1}{m} \sum_{i=1}^{m} x_{ij} \tag{1}$$

The deviation of each element from time mean is written as Δx_{ij}.

Step 3. The similarity criterion ρ_{ik} between i-th and k-th boxes ($1 \leq k \leq n$) is calculated as

$$\rho_{ik} = \frac{d_{ik}^+ - d_{ik}^-}{d_{ik}^+ + d_{ik}^-} \tag{2}$$

where d_{ik}^+ and d_{ik}^- are number of concordant and discordant signs between Δx_{ij} and Δx_{kj} in i-th and k-th boxes time series, $d_{ik}^+ + d_{ik}^- = m$. Corresponding matrix of similarity is square and symmetric i.e. $\rho_{ik} = \rho_{ki}$ and $\rho_{ii} = 1$.

Step 4. The most informative box (etalon-box) is selected from highest average value $\bar{\rho}_i^{\max}$ calculated as

$$\bar{\rho}_i = \frac{1}{n} \sum_{k=1}^{n} \rho_{ik} \tag{3}$$

The etalon-box represents first class. The boxes belong to the first class when $\rho_{ik} \geq \bar{\rho}_i^{\max}$.

Step 5. The remaining boxes are divided into the following classes with corresponding etalons.

Step 6. The final selection of boxes in each class is carried out calculating absolute deviation (AD) of each box from the etalon box and mean absolute deviation (MAD) of concentration value from etalon box in whole class. If the number of cases when AD > MAD exceeds half the number of members of the time series, then this box is excluded from the selected class.

Step 7. The Steps 3–6 are repeated for excluded boxes.

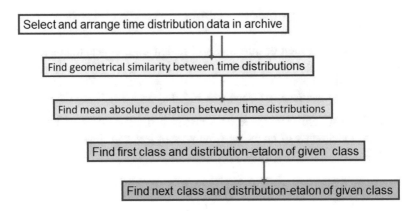

Fig. 1. "Etalon" method classification flow chart.

4   Results of Classification

4.1   Black Sea

The Black Sea is a deep, semi-enclosed marine basin connected with the Mediterranean Sea via the Turkish Straits system (Bosphorus, Sea of Marmara, and Dardanelles). It is also connected with the small, shallow Sea of Azov via the Kerch Strait. A horizontal circulation in the sea consists of two cyclonic (western and eastern) gyres. A box system was built for the Black Sea by using bathymetry data from the Copernicus Marine Environment Monitoring Service [6]. The water column in each box is divided vertically into up to four layers according to water depth, 0–25, 25–100, 100–600 m, and from 600 m to the bottom; the number of layers in each box depends on the total water depth in that box. Advective and diffusional water fluxes between boxes were calculated by averaging three-dimensional currents in the reanalysis data [6] over 10 years (2006–2015). The simulation of the dispersion and fate of ^{137}Cs was carried out for the period 1945–2020. Main sources of ^{137}Cs such as global fallout from atmospheric nuclear weapons tests, atmospheric deposition in result of Chernobyl accident and river runoff from terrestrial areas were considered. Details of model setup and results of simulations are presented in [7].

The results of classification are shown in Fig. 2, where four regions were distinguished: Western, Central and Eastern Black Sea, Northwestern shelf beside the Azov Sea. For each area the etalon-distribution, i.e. temporal distribution that was representative for allocated areas was determined. This distribution was compared with all available measurement data from each area. Comparison of measurements of ^{137}Cs concentration with etalon-distribution for Western Black Sea (box 33), Eastern Black Sea (box 43), Central Black Sea (box 41), and North-Western Black Sea (box 48) is given in Fig. 3.

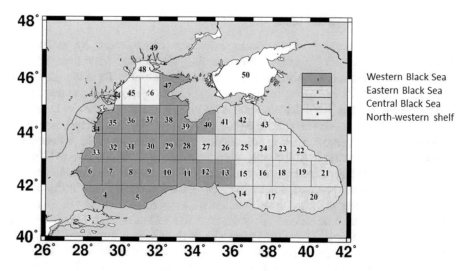

Fig. 2. Regions allocated in the Black Sea as a result of the classification.

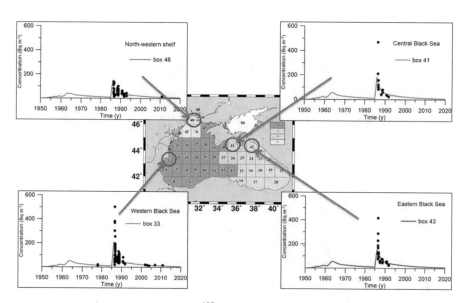

Fig. 3. Comparison of measurements of ^{137}Cs concentration and etalon-distribution for region 1 (box 33), region 2 (box 43), region 3 (box 41), and region 4 (box 48) in the Black Sea.

As seen on the figure, the differences between allocated areas were caused by the fact that the main part of atmospheric deposition after Chernobyl accident fell on the Western Black Sea. In result, in these areas the higher concentration of ^{137}Cs took place, which then was transported by currents of two main gyres. The Central Black Sea region (see Fig. 2) is intermediate between Eastern and Western Black Sea regions, which are placed in the corresponding gyres.

4.2   Northeastern Atlantic Shelf

Another important part of World Ocean essentially contaminated by radioactive materials is northeastern Atlantic Shelf including North Sea, Irish Sea, English Channel, Biscay Bay and adjacent ocean areas. Main sources of ^{137}Cs in this area were global deposition after nuclear weapon testing and Chernobyl accident and due to the routine discharges from the Sellafield and La Hague reprocessing plants. Volume and average depth for each box was calculated using the bathymetry data. Deep boxes were subdivided on three vertical layers to describe the vertical structure of the radioactivity transport in the upper layer (0–100 m), intermediate layer (100–500 m), and deeper layer (>500 m). Details of model setup and results of simulations are presented in [4].

The simulation of the dispersion and fate of ^{137}Cs was carried out for the period 1945–2020. According to the results of simulations, temporal distributions in boxes are extremely different (several orders of magnitude) due to release of ^{137}Cs from Sellafield reprocessing plant into the Irish Sea. The results of classification are shown in Fig. 4, where 11 areas were distinguished. For each area the etalon-distribution, i.e. temporal distribution that was representative for allocated areas was determined. This distribution was compared with all available measurement data in Figs. 5 and 6.

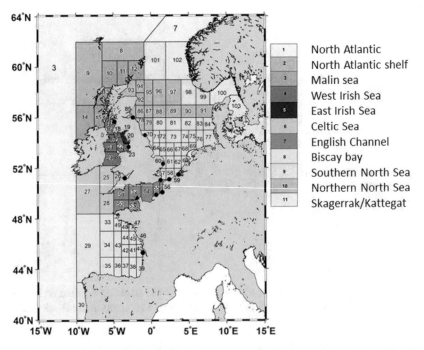

Fig. 4. Regions allocated in the northeastern Atlantic Shelf as a result of the classification.

Figure 5 shows comparison of measurements of ^{137}Cs concentration with etalon-distribution for two areas (Irish Sea and English Channel) where reprocessing plants (Sellafield and La Hague) placed in boxes 20 and 52, respectively. As seen in figure, the release of activity from Sellafield resulted in large gradient of ^{137}Cs concentration even in the Irish Sea. Therefore, the Irish Sea was divided into two areas, in which concentration of ^{137}Cs differed by an order of magnitude. Notice that temporal evolution and magnitude of ^{137}Cs concentration in the English Channel differs from the Irish Sea.

The comparison of measurements of ^{137}Cs concentration with etalon-distribution for northeast Atlantic shelf around the Ireland and Great Britain and Celtic Sea is given in Fig. 6. As seen in figure, the transport of radionuclide through North Passage from

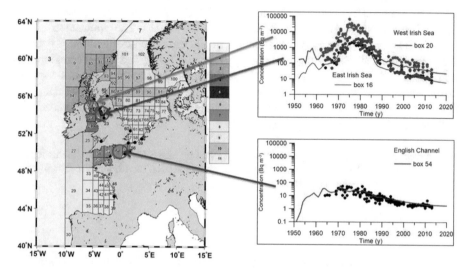

Fig. 5. Comparison of measurements of ^{137}Cs concentration and etalon-distribution for region 4 (box 20) and region 7 (box 54) in the northeastern Atlantic Shelf.

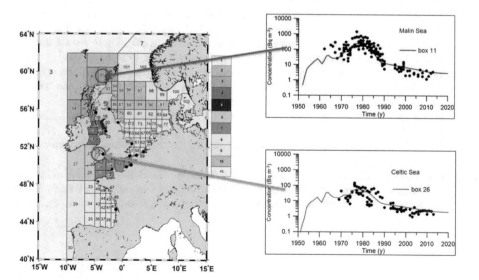

Fig. 6. Comparison of measurements of ^{137}Cs concentration and etalon-distribution for region 3 (box 11) and region 6 (box 26) in the Northwestern European Shelf.

Irish Sea dominates over transport through the Celtic Sea. Then a contaminated water flow is divided on two flows: flow to south along the Great Britain coast and plume propagating to the Norwegian Sea.

5   Conclusions

The "etalon" method was used for classification of the time dependent concentrations of ^{137}Cs calculated by compartment model POSEIDON-R for two marine basins: Black Sea and northeastern Atlantic shelf, for period 1945–2020. The application of this method provides representative distribution of the radionuclide monitoring data for these areas and allows to distinguish 4 regions for Black Sea and 11 regions for northeastern Atlantic shelf. The most representative for allocated regions "etalon" boxes were also determined and compared with all available measurement data. The developed approach can also be used for selection of areas for long-term monitoring of radioactivity levels as most representative locations.

Acknowledgments. This work was partially supported by IAEA CRP K41017 "Behavior and Effects of Natural and Anthropogenic Radionuclides in the Marine Environment and their use as Tracers for Oceanography Studies".

References

1. International Atomic Energy Agency: Sources of radioactivity in the marine environment and their relative contribution to overall dose assessment from marine radioactivity (MARDOS), IAEA-TECDOC-838, IAEA, Vienna, 54 p. (1995)
2. International Atomic Energy Agency: Worldwide marine radioactivity studies (WOMARS). Radionuclide levels in oceans and seas, IAEA-TECDOC-1429, IAEA, Vienna, 187 p. (2005)
3. Lepicard, S., Heling, R., Maderich, V.: POSEIDON/RODOS model for radiological assessment of marine environment after accidental releases: application to coastal areas of the Baltic, Black and North seas. J. Environ. Radioact. 72(1–2), 153–161 (2004)
4. Maderich, V., Bezhenar, R., Tateda, Y., Aoyama, M., Tsumune, D., Jung, K.T., de With, G.: The POSEIDON-R compartment model for the prediction of transport and fate of radionuclides in the marine environment. MethodsX 5, 1251–1266 (2018)
5. Martazinova, V.: The classification of synoptic patterns by method of analogs. J. Environ. Eng. 7, 61–65 (2005)
6. CMEMS, Copernicus Marine Environment Monitoring Service. http://marine.copernicus.eu/. Accessed 26 Apr 2019
7. Maderich, V., Bezhenar, R., Tateda, Y., Aoyama, M., Tsumune, D.: Similarities and differences of ^{137}Cs distributions in the marine environments of the Baltic and Black seas and off the Fukushima Dai-ichi nuclear power plant in model assessments. Mar. Pollut. Bull. 135, 895–906 (2018)

Discrimination of Lithological Types of the Runovshchyna Area for Alpha and Beta Activity

Serhiy Vyzhva[1], Oleksandr Shabatura[1(✉)], Marina Mizernaya[2],
Viktor Onyshchuk[1], and Ivan Onyshchuk[1]

[1] Institute of Geology, Taras Shevchenko National University of Kyiv,
90 Vasylkivska str., Kyiv 03022, Ukraine
dard@ukr.net
[2] D. Serikbayev East Kazakhstan State Technical University,
Ust-Kamenogorsk, Republic of Kazakhstan
mizernaya58@bk.ru

Abstract. The specific alpha and beta activity of the sedimentary rocks of the Runovshchyna area are characterized by statistically significant correlation with the content of uranium and K_2O, which are very similar to those of Volyn-Podillya and Mauritania. The use of alpha-beta radiometry data for direct lithological discrimination is impossible because there is a overlapping of the ranges of parameters in different lithological groups of rocks. For this purpose, it is proposed that linear discriminatory functions for distinguishing clay and sandstone groups. The conducted statistical simulation showed the effectiveness of the allocation of sandstone groups (96%), while for clay only 20%. The low percentage of clay classification is clearly related to the multicomponent of their radionuclide composition. In clay, besides uranium and K-40, there are other alpha and beta emitters (for example from the family of thorium-232). However, a group of incorrectly classified lithological objects is of interest to oil and gas geology as a source of information for the reconstruction of formation and migration of hydrocarbons. An increase in the number of discriminating features of a classification model (if data on the chemical composition of rocks are included) improves the efficiency of distinguishing rules up to 100%.

Keywords: Discriminate model · Alpha-beta radiometry · Runovshchyna area

1 Introduction

1.1 A Subsection Sample

Radioactive properties of rocks have valuable information about the composition and properties of rocks. Most of the natural radioactive elements are related to uranium, actinium and thorium families, which contain, respectively, 20, 15 and 13 genetically related radioactive and stable isotopes. In the uranium series, there are 10 α-, 11 β- and 11 γ-gamma emitters; in actinium row - 10 α-, 7 β- and 6 γ-emitters; in the thorium row - 8 α-, 6 β- and 6 γ-emitters [1]. Each family of radioactive elements has its excellent

A. Palagin et al. (Eds.): MODS 2019, AISC 1019, pp. 21–28, 2020.
https://doi.org/10.1007/978-3-030-25741-5_3

proportions in the activity of each kind of radiation. Taking into account the occurrence of radioactive isotopes in the lithosphere, the intensity of their decay, the greatest influence on the geological processes is caused by uranium, thorium and potassium-40. Actually, the amount of potassium, uranium and thorium minerals, uranium and thorium-containing ones, as well as adsorbed radioactive elements is decisive in the radioactivity of sedimentary rocks.

Determining the content of uranium, thorium and potassium is possible by various methods, but they are usually quite complex and costly. In logging wells and laboratory studies, gamma and neutron methods are usually used to record relevant radiation [4]. Therefore, the use of the alpha-beta method, as more expressive and simpler in the laboratory studies of sedimentary rocks will allow to perform a qualitative and, if possible, quantitative assessment of the radioactivity of different lithological types of rocks.

2   Modelling

2.1   Measuring

In the laboratory of nuclear geophysics of the Department of Geophysics, the Institute of Geology of the Kyiv University, a series of alpha and beta-analysis of powder samples was carried out with the help of a laboratory low-background device of UMF-2000, which provides high-precision determination of the specific alpha (α, Bq/kg) and beta (β, Bq/kg) activity of samples of rocks, soils and dry residue water samples [4] (Table 1).

Table 1. Correlations between the content of uranium (C_U) and K_2O (C_{K2O}) and specific alpha and beta activity in sedimentary rocks of different regions.

Region	Correlation	R^2	Volume	Source
Volyn-Podillya	$C_{K2O} = 1{,}6 \cdot 10^{-3}\beta + 5{,}5 \cdot 10^{-2}$	0,916	107	[4]
Runovshchyna area	$C_{K2O} = 1{,}6 \cdot 10^{-3}\beta + 6{,}6 \cdot 10^{-2}$	0,992	191	
Volyn-Podillya	$C_U = 3 \cdot 10^{-7}\alpha + 5 \cdot 10^{-5}$	0,863	107	[4]
Runovshchyna area	$C_U = 3 \cdot 10^{-3}\alpha + 0{,}5032$	0,999	191	
OumDheroua, Mauritania	$C_U = 7{,}7 \cdot 10^{-3}\alpha + 1{,}0396$	0,928	87	[5]

The given correlation dependences were high to evaluate the content of K2O in different-age formations and lithological types of Volyn-Podillya [4]. Clays have an average higher content of uranium and, accordingly, α due to its concentration by sorption on clay particles, the formation of quaternary uranium or the formation of solid solutions under redox.

Sandy rocks may contain uranium minerals, often as an isomorphous admixture to heavy fraction minerals. Compared with alpha activity, beta-activity has a significantly better discriminative ability, which allows to distinguish between rocks of different lithological composition.

Instead, the ranges of values α- and β-activity of sedimentary rocks in Runovshchyna area which makes it impossible to distinguish them easily using these parameters (Fig. 1).

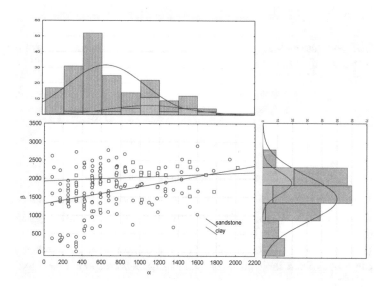

Fig. 1. Alpha and beta activity of sandstone and clay of Runovshchyna area.

Two groups of objects (clay and sandstone) are presented in the coordinate system of two signs - α- and β-activity. The density of the probability of signs has a two-modal character. If it use radiometric features separately, then the confident distinction will not be executed. Classification error is proportional to the density of overlap area of variation. Therefore, the task of constructing a decision rule for discrimination of two lithological groups of objects is based on mathematical statistics methods. For these purposes, it is proposed to use a decision rule with the linear discrimination functions (LDF). LDF include two signs (α- and β-activity) to divide the primary sample into two classes I and II (sandstone and clay).

2.2 Linear Discriminatory Functions

LDF is used to construct a decision rule for the discrimination of two groups of objects by a fixed set of signs [4]. In general, the equation has the form:

$$D_i = \lambda_1 x_{i1} + \lambda_2 x_{i2} + \ldots + \lambda_m x_{im} = \sum_{k=1}^{m} \lambda_k x_{ik} \qquad (1)$$

where x_{ik} - the value of the k-th sign for the i-th object; $\lambda_1, \lambda_2, \ldots \lambda_m$ - coefficients of LDF, D_i - a discriminator:

If $D_i > D_0$ then the object refers to the first group, the opposite case - to the second one. The boundaries between a group of objects is a straight line whose equation (for two signs) is:

$$x_2 = D_0/\lambda_2 - \lambda_1 x_1/\lambda_2 \tag{2}$$

The LDF coefficients that perform the discrimination of objects into groups are the roots of a system of m linear equations.

$$
\begin{aligned}
\lambda_1 k_{11} + \lambda_2 k_{12} + \ldots + \lambda_m k_{im} &= q_1 \\
\lambda_1 k_{21} + \lambda_2 k_{22} + \ldots + \lambda_m k_{2m} &= q_2 \\
&\ldots \\
\lambda_1 k_{m1} + \lambda_2 k_{m2} + \ldots + \lambda_m k_{nm} &= q_m
\end{aligned}
\tag{3}
$$

where $k_{ij} = \left(\frac{N_I k_{ij}^I + N_{II} k_{ij}^{II}}{N_I + N_{II} - 2}\right)$ - the elements of the combined covariance matrix I, which are calculated by the corresponding values k_{ij}^I, k_{ij}^{II} of the group matrices K_I and K_{II}; N_I, N_{II} - group sizes; $q_i = \bar{a}_i - \bar{b}_i$ - the difference between the mean and i-th signs in groups I and II [2].

The threshold value of the discriminator is determined by substituting the arithmetic mean values from the centers of groups of the corresponding attributes in the LDF equations:

$$D_0 = \sum_{k=1}^{m} \lambda_k \left(\frac{\bar{a}_i - \bar{b}}{2}\right). \tag{4}$$

2.3   Results of Classification

The calculations carried out in the program Statistica 7.0 and received the following LDF:

$$\text{Sandstone} = 0,00228\alpha + 0,00419\beta - 1,28126 \tag{5}$$

$$\text{Clays} = 0.00477\alpha + 0.00496\beta - 9.51348 \tag{6}$$

Lambda $\alpha = 0,9931963$ is greater than lambda $\beta = 0,853632$, which suggests the advantage of incorporating such a feature into a discriminatory model as α -activity compared to β-activity. The classification results are presented in Fig. 2 and Table 2.

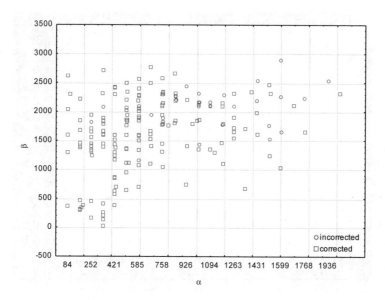

Fig. 2. Results of the classification of Runovshchyna area in lithological type using radiometric data.

Table 2. Classification matrix (Runovschyna area, rows: observed classifications, columns: predicted classifications).

	Percent correct classified	Sandstone	Clay
Sandstone	95,57	151	7
Clay	20	24	6
Total	83,51	175	13

The decision rule on the basis of LDF can be improved and the use of a model in addition to radiometric data, signs of chemical composition, which increases the effectiveness of discrimination up to 100% (Table 3).

Table 3. Classification matrix (Runovschyna area, rows: observed classifications, columns: predicted classifications).

	Percent correct classified	Sandstone	Clay
Sandstone	100	60	0
Clay	100	0	4
Total	100	60	4

In the expanded model, the most informative are not radiometric indicators, and such indicators include titanium dioxide, manganese, magnesium and water content (Fig. 3).

N=64	Wilks' Lambda	Partial Lambda	F-remove (1,48)	p-level	Toler.	1-Toler. (R-Sqr.)
α	0,166593	0,996179	0,18412	0,669778	0,735508	0,264492
β	0,177547	0,934720	3,35226	0,073324	0,414308	0,585692
SiO2	0,169830	0,977194	1,12023	0,295167	0,024532	0,975468
TiO2	0,184336	0,900295	5,31588	0,025497	0,128197	0,871803
Al2O3	0,178672	0,928833	3,67774	0,061104	0,152430	0,847571
Fe2O3	0,166822	0,994812	0,25031	0,619147	0,204995	0,795005
MnO	0,185086	0,896647	5,53277	0,022813	0,061656	0,938344
MgO	0,208608	0,795544	12,33610	0,000979	0,057592	0,942408
CaO	0,171608	0,967067	1,63461	0,207214	0,038413	0,961587
Na2O	0,166011	0,999671	0,01580	0,900482	0,102555	0,897445
K2O	0,176950	0,937876	3,17945	0,080894	0,108627	0,891373
P2O5	0,174889	0,948929	2,58336	0,114552	0,171048	0,828952
S	0,168804	0,983135	0,82340	0,368722	0,140663	0,859337
Cl	0,176605	0,939706	3,07978	0,085652	0,243305	0,756695
H2O	0,219190	0,757135	15,39686	0,000277	0,323177	0,676823

Fig. 3. Discriminate function analysis summary of expanded model.

The obtained LDF in the model of alpha-beta radiometry indicate a rather effective distinction between sandstone and much worse clay. Incorrected classified objects are mainly associated with clay as products of high degree of chemical transformation of the rock. It was affected with high Fe-content and decrease the content of silicates. So can observe growth of petrochemical modules: hydrolyzate module, HM and alumina module AM (Fig. 4).

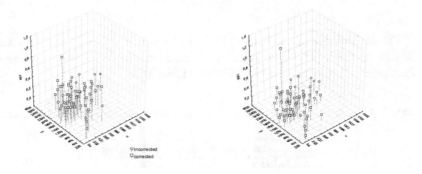

Fig. 4. The groups are correctly and incorrectly classified lythotype of the Runovshchyna area under AM (a) and HM (b) in the coordinates of α- and β-activity

Clay deposits have on average more potassium and α-activity due to its sorption of K_2O on clay particles (Fig. 4(a)). Specific α-activity shows a clear differentiation to HM. Highly active deposits of the Runovshchyna area are associated with higher values of HM, respectively, they can be combined with the conditions of strong and deeper weathering. To distinguish between lithotypes, this tendency was critical, since samples with the highest α-activity are clay, and low-activity - sandstone. The smaller the HM, the higher the maturity of the sedimentary rock, and hence the lower specific α-activity.

It is known that the relationship between concentrations of uranium, thorium and potassium serve as important signs of conditions for the formation of sedimentation, which may be an indicator of the potential of oil and gas accumulation [4]. Symptomatically, that objects with low parameters of potassium modulus PM occurs into a group of incorrectly classified rock (Fig. 5). PM is used by oil geologists to predict the oil and gas bearing strata, since it carries information on the distribution of the genetic type of clay, which is associated with the conditions of oil and gas production. The impact of the PM is not so obvious, because sandy rocks always contain potassium. And their increased radioactivity is associated with potassium polymorphic varieties. The minimum content of potassium is noted in well-sorted marine, mostly quartz sandstone; elevated - for clay varieties of sandstones, as well as sandstones with organic impurities.

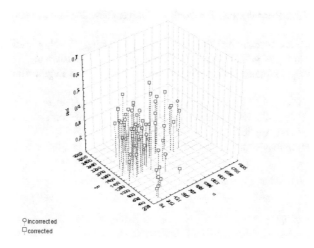

○ incorrected
□ corrected

Fig. 5. The groups of correctly and incorrectly classified lythotypes of the Runovshchyna area under PM the coordinates of α- and β-activity.

Since the clay rocks generate micro-oil and gas, which are further depressed and localized in other rocks, depending on the mineral composition, it can be assumed that the recovery of oil or gas up the profile [3].

Upon the possibility of reconstruction of the formation conditions, several types of rocks are identified which correspond to different conditions of oil and gas formation: α = 1000–1500 (Bq/kg), β = 1800–2000 and α = 300–1100, β = 600–1400 - conditions of strong and deep weathering, which can be connected to the maternal layer of the kerogen; and, the other group with α = 100–700, β = 200–1800 - mature sedimentary rocks, as a situation of migration and accumulation of hydrocarbons.

2.4   Conclusions

The specific alpha and beta activity of the sedimentary rocks of the Runovshchyna area show statistically significant correlation with the content of uranium and K_2O. The use of alpha-beta radiometry data for direct lithological discrimination is impossible because there is a overlapping of the ranges of parameters in different lithological groups of rocks. The use of linear discriminatory functions for selection of clay and sandstone groups is proposed. The effectiveness of the allocation of sandstone groups is quite high and is 96%, clay only 20%. The low percentage of clay classification is clearly related to the multicomponent of their radionuclide composition. In which, besides uranium and K-40, there are other alpha and beta emitters, for example from the family of thorium. But a group of incorrectly classified lithological objects is of interest to oil and gas geology as a source of information for the reconstruction of the formation and migration of hydrocarbons.

References

1. Kobranova, V.N.: Petrophysics. Textbook for Universities. 2nd edn., Nedra, 392 p. (1986). [in Russian]
2. Shcheglov, V.I.: Mathematical methods of modeling in geology. Electronic edition/South-Ros. State Techn. Un-t, 197 p. YURGRT, Novocherkask (2012). [in Russian]
3. Sklyarov, E.V. (ed.): Interpretation of geochemical data. Internet Engineering, 288 p. (2001). [in Russian]
4. Vyzhva, S.A., Onyshchuk, D.I., Onyshchuk, I.I. Radiometric parameters of reservoir-rocks of hydrocarbon Volyno-Podillya prospects. In: 11th International Conference on Geoinformatics: Theoretical and Applied Aspects, Geoinformatics, Kiev (2012)
5. http://www.geol.univ.kiev.ua/docs/conf/conf_univ_apr_2019_program.pdf. [in Ukrainian]

The Model of the Pollution Spread in the Cascades of Ponds Within the Protected Areas

Vitaliy I. Zatserkovnyi⬤, Kateryna A. Kazantseva$^{(\boxtimes)}$⬤,
and Ludmila V. Plichko⬤

Institute of Geology, Taras Shevchenko National University of Kyiv,
90 Vasilkivska Str., Kyiv 03022, Ukraine
vitallii.zatsekovnyi@gmail.com, djanaia@ukr.net

Abstract. The paper presents the developed model of the pollution spread in the cascades of ponds in the Holosiivskyi National Natural Park. The mathematical model is based on graph theory and dynamic equations. Lake cascades are shown in the form of oriented graphs which form a single hydrological network of a complex hierarchy. Therefore, from the point of view of mathematics, they form a tree that is a strictly hierarchical orgraph whose vertices are loaded using a dynamic equation. As a result, the pollution of ponds is modeled as a dynamic not statistical system and is not empirically defined as a constant (Lotka-Volterra model). In this paper, pollution is a dynamic that occurs in a geo-ecosystem and fluctuates between pollution and recovery approaching equilibrium. It means that the dynamic equation tries to achieve equilibrium. The task of loading graph vertices is based on the Lotka-Volterra equations with constraints that enable to assess the behavior of the environment, which is in constant progress according to the pollution. Ponds in a cascade are either being polluted or self-healing after pollution. The authors propose to consider pollution as a dynamic process that consists of pollution and recovery, unlike whose who consider this indicator as an empirically defined constant.

Keywords: Mathematical modeling · Pollution · Graph theory ·
Dynamic equations · Geo-ecology

1 Introduction

Although a number of issues on modeling the spread of pollution in water bodies have been investigated [1–5], they are mainly represented by probabilistic or statistical models that inadequately describe the processes occurring in them. Much has been done in the field of researching the pollution spread in the cascades of ponds within protected areas, but undoubtedly much remains to be done. Since natural systems are stable in time, throughout their existence, when a polluter is introduced, they seek to counteract this process by self-purification.

Section 2, There is defined and mathematized a task of modeling the process of polluting and restoring the environment. This dynamic model seeks to achieve equilibrium. In Sect. 3 there are shown the main results of two permissible lim - 1 - when

© Springer Nature Switzerland AG 2020
A. Palagin et al. (Eds.): MODS 2019, AISC 1019, pp. 29–36, 2020.
https://doi.org/10.1007/978-3-030-25741-5_4

the environment is modified under the action of a pollutant and is unable to recover; 2 - when the effect of pollution is not significant and the environment has not changed. Section 4 presents the main result of the model which takes into account the influence of the environment on pollution. Thus, the solution of the model, where the influence of the environment on pollution is not considered, may be used to estimate the decoupling of a more accurate model taking into account the influence of the environment. At the same time, the pollution in the first model evaluates the pollution in the second one from above, and the medium in the first model estimates the medium in the second model from below. However, a general solution of a dynamic equation, such as Lotka-Volterra which describes the dynamics occurring in the geo-ecosystems under the influence of pollution, is presented.

2 Introduction

Let us assume that $u(t,x,y)(v(t,x,y))$ is the concentration of the pollutant and the power of medium recovery at time t at the point with x, y coordinates. This task can be calculated using the well-known Lotka-Volterra model.

Accordingly, from a mathematical point of view, this vector may be negative, but since we denote the wind direction by it, we state its inseparability. The movement of the pollutant is described with the diffusion equation.

Therefore, considering the micro-level territory, it is possible to change and, in particular, simplify this equation and neglect the vector of the average annual flow, as at the local level there will be no such significant indicator and this indicator can be equated to 1.

$$u_t = af(x - x_0, y - y_0) - gu + D(uxx + uyy)$$
$$vt = rv(1 - v/K) - Cuv \tag{1}$$

Explanation of the formula: α is the power of the pollution source, $f(x - x_0, y - y_0)$ is a function characterizing the pollution source, (x_0, y_0) – the source coordinates, r, K– a parameter of the logistic equation describing the exchange process in the medium which occurs if there is no pollution; D is the coefficient of pollution diffusion, C_{uv} is a function of the pollutant influence on the medium of a pond, $Auv/(B + u)$ is a function of the influence of the environment on a pollutant.

Since this dynamic system is not trivial, does not have a generally accepted solution and cannot be dealt with the help of existing algorithms, the process of its solving has significant amounts. It is worth noting that the model is a non-linear second-order differential equation with a gamma function and disturbances. As known, these types of equations are considered to be rather complicated for any analytical solution; therefore, in such case, partial numerical solutions are considered.

Bearing in mind the work of Grodzinsky [2] on the disturbance of landscapes, this model should be considered as of two types: when the disturbances in the medium are small, $Auv/(B + u) - $ const $\leq 1|50$, while this equation is simplified, and when $Auv/(B + u)$ is variable and can take values more than 1/50 which leads to significant disturbances, fluctuations. It is also important to note that the use of the expression

$Auv/(B+u)$ enables to present the interrelations between vacationers and the environment, that will permit us to calculate vacationers' impact on the environment, evaluate their interrelations and consider not just a single pollutant but recreational or any anthropogenic activity as the pollution source. It is necessary to evaluate all types of pollution, both actual and latent.

3   Results

If we assume that $k_x = k_y = A = 0$, then the dynamics of pollution with sources at the points $(x_1; y_1); \ldots; (x_k; y_k)$ is described as:

$$\text{References } u_t = \sum_{i=1}^{k} \alpha_i f_i(x - x_i, y - y_i) - gu + D\Delta u,$$
$$u\big|_{t=0} = u_0(x, y),$$
$$\tag{2}$$

where $\Delta u := u_{xx} + u_{yy}$.

The solution to this problem can be found explicitly. Make a replacement $u = e^{-gt}z$

$$z\big|_{t=0} = u_0(x, y) \tag{3}$$

It is known that en the solution for $t > 0$ will give the Poisson formula

$$Z(x, y, t; D, g) = \sum_{i=1}^{k} \int_0^t \int_{\mathbb{R}^2} \frac{\alpha_i f_i(\xi - x_i, \eta - x_i)}{4\pi D(t - \tau)} exp\left\{ g\tau - \frac{(x - \xi)^2 + (y - \eta)^2}{4D(t - \tau)} \right\} d\xi \, d\eta d\tau$$
$$+ \frac{1}{4\pi Dt} \int_{\mathbb{R}^2} u_0(\xi, \eta) exp\left\{ g\tau - \frac{(x - \xi)^2 + (y - \eta)^2}{4Dt} \right\} d\xi d\eta. \tag{4}$$

Thus,

If, like in the work of Bratus [1], the sources of contamination are pointlike, which means $f_i(x; y) = (x; y)$ is the Dirac Delta Function, then at all points $(x; y)\sigma = (x_i; y_i)$ $i = 1; \ldots; k$, and when $t > 0$ the function $u = (x, y, t, D, g)$ is determined by the formula:

$$u(x, y, t; D, g) = \sum_{i=1}^{k} e^{-gt} \int_0^t \frac{\alpha_i e^{g\tau}}{4\pi D(t - \tau)} exp\left\{ -\frac{(x - x_i)^2 + (y - y_i)^2}{4D(t - \tau)} \right\} d\tau$$
$$+ \frac{e^{-gt}}{4\pi Dt} \int_{\mathbb{R}^2} u_0(\xi, \eta) exp\left\{ -\frac{(x - \xi)^2 + (y - \eta)^2}{4Dt} \right\} d\xi d\eta \tag{5}$$

We analyze the expressions that are presented here at points remote from each pollution source at a distance not less than ρ . We have:

$$e^{-gt} \int_0^t \frac{\alpha_i e^{g\tau}}{4\pi D(t - \tau)} exp\left\{ -\frac{(x - x_i)^2 + (y - y_i)^2}{4D(t - \tau)} \right\} d\tau = [\tau = t - s]. \tag{6}$$

If g = 0, then it may be shown that each above-mentioned function will grow to infinity when t → 1 according to a logarithmic law.

Further,

$$\frac{e^{-gt}}{4\pi Dt} \int_{\mathbb{R}^2} u_0(\xi, \eta) exp\left\{ -\frac{(x-\xi)^2 + (y-\eta)^2}{4Dt} \right\} d\xi d\eta \tag{7}$$

$$= \left[\xi = x + 2\sqrt{Dt}\xi_1, \eta = y + 2\sqrt{Dt}\eta_1 \right]$$

$$= \frac{e^{-gt}}{\pi} \int_{\mathbb{R}^2} u_0\left(x + 2\sqrt{Dt}\xi_1, y + 2\sqrt{Dt}\eta_1\right) exp\left\{ -\xi_1^2 - \eta_1^2 \right\} d\xi d\eta$$

$$\leq e^{-gt} \max_{(x,y) \in \mathbb{R}^2} u_0(x, y). \tag{8}$$

Therefore, this function remains bounded, and if $g > 0$, then it tends to 0 when $t \to \infty$. Based on the above-mentioned reasoning, we can conclude:

If $g > 0$, there is a finite boundary at the points remote from each source of pollution at a distance not less than ρ:

$$\lim_{t \to \infty} u(x, y, t) = u_*(x, y) \leq \frac{1}{4\pi e \rho^2 g} \sum_{i=1}^{k} \alpha_i \tag{9}$$

The initial distribution of pollution $u_0(x; y)$ should be determined on the basis of experimental data $u_{i,j}^{\sim} x$ on some grid of points $(\xi_i, \eta_j), i = 1, \ldots, N, j = 1, \ldots, M$. Then $u_0(x; y)$ may be specified, for example, by the interpolating Lagrange polynomial

$$u_0(x, y) = \sum_{n=1}^{N} \sum_{m=1}^{M} u_{n,m}^{\sim} \prod_{i=1, i \neq n}^{N} \prod_{j=1, j \neq}^{M} \frac{(x - \xi_i)(y - \eta_j)}{(\xi_n - \xi_i)(\eta_m - \eta_j)} \tag{10}$$

For numerical calculations the integral in the second term of the expression for u is replaced, for example, with a circle of a sufficiently large radius. A more crude approach is to divide the study area into rectangles R_{ij} with centers at the points (ξ_i, η_j) and assume that $u_0(x; y) = u_{ij}^{\sim}$, when (x; y) belongs to R_{ij}. Then

$$\frac{1}{t} \int_{\mathbb{R}^2} u_0(\xi, \eta) exp\left\{ -\frac{(x-\xi)^2 + (y-\eta)^2}{4Dt} \right\} d\xi d\eta \tag{11}$$

$$\approx \sum_{i=1}^{N} \sum_{J=1}^{M} \frac{u_{i,j}^{\sim}}{t} \int_{R_{ij}} exp\left\{ -\frac{(x-\xi)^2 + (y-\eta)^2}{4Dt} \right\} d\xi d\eta$$

The parameters $\alpha_i, i = 1, \ldots, k$ are determined experimentally: this is the magnitude of the pollution emitted by the first source per unit time. To determine the parameters g and D, it is possible to experimentally find the magnitude of contamination at times $t_1; : : : ; t_s$ at the points of which grid $\left(\xi_i', \eta_j'\right), i = 1, \ldots, N', j = 1, \ldots, M'$. Then the values $u_{ijl}^{\sim}$. Then calculate $u\left(\xi_i', \eta_j', t_l; D, g\right)$ Then $D; g$, is found by minimizing the functional $\sum_{i=1}^{N'} \sum_{j=1}^{M'} \sum_{l=1}^{s} \left[u\left(\xi_i', \eta_j', t_l; D, g\right) - \hat{u}_{ijl}\right]^2 \rightarrow min$ with respect to the variables $D; g$.

If we assume that the function $u(x; y; t)$ is known, then for a recreational environment we obtain the equations:

$$v_t = rv\left(1 - \frac{v}{K}\right) - Cu(x, y, t)v, \tag{12}$$

or

$$\frac{dv}{dt} = [r - Cu(x, y, t)]v - \frac{r}{K}v^2. \tag{13}$$

In the equation above $x; y$ are presented as parameters. We introduce the notation $a(t) := r - Cu(x; y; t), b = r/K$ We obtain the Bernoulli equation:

$$\frac{dv}{dt} = a(t)v - bv^2 \tag{14}$$

It can be integrated into quadratures. We divide both sides of the equation on v^2:

$$\frac{1}{v^2}\frac{dv}{dt} = \frac{a(t)}{v} - b$$

and introduce a new variable:

$$\omega = \frac{1}{v}$$

So,

$$-\frac{d\omega}{dt} = a(t)\omega - b \Leftrightarrow \frac{d\omega}{dt} = -a(t)\omega + b. \tag{15}$$

This is a linear inhomogeneous equation. Its general solution is:

$$\omega = exp\left\{-\int_0^t a(s)ds\right\}\left[c + b\int_0^t exp\left\{\int_0^\tau a(s)ds\right\}d\tau\right], \tag{16}$$

where c has become arbitrary. Considering the initial condition

$$v(0) = v_0, \omega(0) = 1/v_0, c = 1/v_0$$

So, $c = 1 = v_0$, and the solution to the Bernoulli equation that satisfies the initial condition is:

$$v = \frac{v_0 exp\left\{\int_0^t a(s)ds\right\}}{1 + v_0 b \int_0^t exp\left\{\int_0^\tau a(s)ds\right\}d\tau},$$

That is

$$v(x,y,t) = \frac{v_0 exp\left\{rt - C \int_0^t u(x,y,s)ds\right\}}{1 + v_0 b \int_0^t exp\left\{r\tau - C \int_0^\tau u(x,y,s)ds\right\}d\tau} \tag{17}$$

It was shown above that there is a boundary

$$\lim_{t\to\infty} u(x,y,t) = u_*(x,y).$$

If

$$r > Cu, (x,y),$$

then we can calculate the limit $\lim_{t\to\infty} u(x,y,t)$. Lopital rule

$$\lim_{t\to\infty} v(x,y,t) = \lim_{t\to\infty} \frac{v_0 exp\left\{rt - C\int_0^t u(x,y,s)ds\right\}(r - Cu(x,y,t))}{v_0 b exp\left\{rt - C\int_0^t u(x,y,s)ds\right\}} = \frac{r - Cu_*}{b} > 0. \tag{18}$$

So, in this case, the natural environment survives. if

$$r < Cv_*(x,y),$$

That

$$\lim_{t\to\infty} \frac{1}{t} \int_0^t u(x,y,s)ds = u_*(x,y),$$

In this case, we get:

$$\lim_{t\to\infty} exp\left\{rt - C\int_0^t u(x,y,s)ds\right\} = \lim_{t\to\infty} exp\left\{t\left(r - \frac{C}{t}\int_0^t u(x,y,s)ds\right)\right\} = 0 \tag{19}$$

and $\lim_{t\to\infty} v(x,y,t) = 0$ In this case, the medium no survives.

Consider a model that takes into account the factor of the influence of the recreation zone on pollution:

$$u_t = \sum_{i=1}^{k} \alpha_i f_i(x - x_i, y - y_i) - gu - \frac{Auv}{B+u} + D\Delta u,$$

$$v_t = rv\left(1 - \frac{v}{K}\right) - Cuv,$$

$$u|_{t=0} = u_0(x,y), v|_{t=0} = v_0(x,y).$$

The solution of this problem is denoted by $u_1(x,y,t)$, $v_1(x,y,t)$. Then the difference $U = u - u_1$ solution of the problem, which we considered above, satisfies the following expression

$$U_t = \frac{Au_1 v_1}{B + u_1} - gU + D\Delta U$$

$$U|_{t=0} = 0$$

So,

$$U(x,y,t) = e^{-gt} \int_0^t \int \int \frac{e^{g\tau}}{4\pi D(t-\tau)} \frac{Au_1(x,y,\tau)v_1(x,y,\tau)}{B + u_1(x,y,\tau)}$$

$$\times \exp\left\{ -\frac{(x-\xi)^2 + (y-\eta)^2}{4D(t-\tau)} \right\} d\xi d\eta d\tau > 0 \tag{20}$$

Thus: $u(x,y,t) > u_1(x,y,t), t > 0$
Then

$$f_1(x,y,t,v) := rv\left(1 - \frac{v}{K}\right) - Cu_1(x,y,t,)v > rv\left(1 - \frac{v}{K}\right)$$

$$- Cu(x,y,t,) := f(x,y,t,v) \tag{21}$$

Since $v(x; y; t)$ та $v_1(x; y; t)$ are, respectively, solutions of problems And $v_t = f_1(x,y,t,u)$, $v|_{t=0} = v_0(x,y)$ by the well-known comparison theorem

$$v_1 = (x,y,t,u) > v(x,y,t), t > 0 \tag{22}$$

Consequently, we have considered the general solution of the problem 1, special cases directly depend on the indicator t. As a result, taking different time intervals, we can study the current state of the system and model its behavior in the future. However, this model makes it possible to investigate the pollution dynamics of geo-ecosystems with sufficiently small oscillations of the system. Since, both theoretically and empirically, it has been proved that the significant splashes of pollution will lead to the significant changes in the system bringing it to a new qualitative level, we limit $Auv/(B+u)$ making it as small as possible.

4 Conclusion

The article presents the model of the pollutant spreading in a cascade of reservoirs (ponds). This model makes it possible to assess pollution of the cascade of reservoirs as a complex dynamic model that varies from the state of pollution to a state of self-healing. The interconnection function itself tends to get closer to equilibrium, which will lead to the stabilization of the system. The task is based on the "prey and predator" task with the modified edge points. The solution is presented in general.

References

1. Bratus, A., Novozhilov, A., Platonov, A.: Dynamical systems and models in biology. Fizmatlit (in Russian) (2010)
2. Grodzinsky, D., Svidzinskaya, D., Golotka, D., Tishayeva, A.: FREEWAT international water planning and management program: goals and implementation process. Ukr. Geogr. J. 4(4), 17–21 (2016)
3. Martynyuk, A.A.: Elements of Time Scale. Springer, Switzerland (2016)
4. Zomorrodi, A.R., Segrè, D.: Synthetic ecology of microbes: mathematical models and applications. J. Mol. Biol. 428(5), 837–861 (2016). Part B
5. Ramsay, J., Hooker, G.: Dynamic Data Analysis Modeling Data with Differential Equations. Springer, Switzerland (2017)

Mathematical Modeling and Simulation
of Systems in Manufacturing

Using Mathematical, Experimental and Statistical Modeling to Predict the Lubricant Layer Thickness in Tribosystems

Nikolay Dmytrychenko⬤, Viktoriia Khrutba$^{(\boxtimes)}$⬤,
Anatoliy Savchuk⬤, and Andrii Hlukhonets⬤

National Transport University,
Omelianovycha-Pavlenko st. 1, Kyiv 01010, Ukraine
dmitrichenko@ntu.ua, Viktoriia.Khrutba@gmail.com,
tolik_savchuk@bigmir.net, hanti@i.ua

Abstract. Main performance characteristics of any tribosystem depend on the correct selection, quality and properties of structural and lubricating materials, surface roughness (in particular, volumetric spatial configuration) and macro-geometric indicators. In the modern world of tough competition there is a need to reduce the time and costs of laboratory research and replace them with a computational experiment using the developed mathematical models. The aim of this research is to develop a multivariable regression model depending on the actual thickness of the lubricant from the factors identified in local contact. It will enable to predict the properties of structural and lubricating materials in tribosystems to improve performance in modern machines and mechanisms. The investigated model factors were chosen: the range of change is fast varying from 0 to 1.8 m/s; the volumetric oil temperature during the experiment varied from 293.1 K to 343.1 K; the contact load was 251.5 MPa. Initial parameter is thickness of the lubricating layer in contact determined by the method of optical interference. The experimental results were processed by the least squares method to obtain adequate mathematical model of the second order depending on the thickness of the lubricant from changing factors. The correlation coefficient is an average of 0.98. The relative error of the predicted values of the output variable does not exceed 5%. The resulting dependence to predict the properties of the lubricant from the technical parameters were specified in local contact.

Keywords: Friction · Lubricant · Mathematical modeling · Lubricating film thickness

© Springer Nature Switzerland AG 2020
A. Palagin et al. (Eds.): MODS 2019, AISC 1019, pp. 39–49, 2020.
https://doi.org/10.1007/978-3-030-25741-5_5

1 Background of Mathematical, Experimental and Statistical Modeling to Predict the Lubricant Layer Thickness in Tribosystems

1.1 Problem Statement and the Purpose of Research

The rapid growth of engineering industry requires new technical solutions and creating enterprises which can realize the complete production cycle including not only assembly machines, but manufacturing all units, parts and their quality control. Modern machines and mechanisms are set tribosystems. The force of friction, tribosurface wear rate, noise, vibration - these factors affect the performance and, as a result the service life of the tribo-units. The main performance characteristics of any tribosystems depend on the proper selection, quality and properties of structural and lubricating materials, surface roughness (including volumetric spatial configuration) and macrogeometrical indicators. All the above-mentioned problems should be solved at the tribo-units construct stage on the necessary design calculations (which, unfortunately, are not accurate and complete in tribology) and methods and means of laboratory research. In the modern world of tough competition there is a need to reduce the time and costs of laboratory research and replace them with a computational experiment using the developed mathematical models. Consequently, the creation of adequate mathematical models enabling to predict the properties of structural and lubricating materials in tribosystems to improve performance indicators in modern machines and mechanism is an actual scientific problem.

The aim of this research is to develop a multivariable regression model depending on the actual thickness of the lubricant from the factors identified in local contact.

1.2 The Analysis of Recent Research and Publications

A lot of current national and international research is devoted to the issues of selection, quality, properties of structural and lubricating materials, surface roughness and macrogeometrical indicators.

The works of Mikosianchyk [1] and Savchuk [2] present the research on the lubricating properties of lubricants and the impact of operational and rheological factors in different conditions of tribosystems. The authors determined that the operational indicators of vehicle units had a huge effect on the formation dynamics of the lubricating layer thickness, which is the main characteristic feature of lubrication rate. Lubricating layer thickness causes the lubricating ability of modern lubricants and therefore affects the wear resistance of the friction surfaces. Determining the medium and minimum thickness of the lubricant layer is important for the practical side of the elastodordinomic problem for local contact. The value of the film in contact is an important practical side for durability of the unit components and mechanisms.

The works including a part of mathematical modeling are of particular interest:

Friction of metal-matrix self-lubricating composites: Relationships among lubricant content, lubricating film coverage, and friction coefficient [3], the prediction of wear-loss amounts of ferro-alloy coating using different machine learning algorithms [4], a model of contact friction based on extreme value statistics [5], assessing the workability of greased bearings after long-term storage [6].

The issue of constructing regression models according to experimental data and determining their adequacy at acceptable risk and level of error is devoted to work [7, 8]. Solutions of scientific and technological problems in experimental statistic models with the help of visualization tools ("quasi-factor" curves on a square, surface of response on cube, etc.) are considered in [9, 10]. Efficiency is proved of this methods for reserche of composite building materials [11], chemical-technological processes and systems [12, 13], individual processes in mechanical engineering [14] and many others. However, the experience of using experimental-statistical simulation for predicting of the lubricat layer thickness, depending on the conditions of local contacts, the properties of structural and lubricant materials in tribosystems is absent.

1.3   Materials and Methods of Research

To develop mathematical models enabling to predict lubricant layer thickness, depending on the conditions of local contacts, the properties of structural and lubricating materials in tribosystems to improve performance in modern machines and mechanisms, experimental and statistical modeling approaches were used [7, 8]. The task of experimental and statistical modeling is to establish a relationship between input parameters - output parameters and factors - quality indicators of the system and determine the levels of factors optimizing the output system parameters. The construction of such physical systems models does not require system knowledge or any other information about the processes occurring inside the country. Consequently, empirical models work on the principle of a black box and do not contain a physically grounded function connecting data at the input of the system with the parameter characterizing its state. The harmonization of input and output parameters when creating such models is based on the statistical processing of experimental data, on the methods of probability theory and mathematical statistics.

Methods of experimental research include practical realization of working conditions of refractory rolling bearings for determining the actual lubricating layer thickness in contact with friction pairs. The lubricant layer thickness in contact was determined by optical interference method.

The results of experiment and calculation coefficients of models were processed by least squares method (LSM). The essence of approximation of the LSM provides for the table data obtained in the experiment, the search for analytic dependence, the sum of squares of deviations from the table data at all unit points would be minimal.

The general formula of LSM in matrix form is [15]:

$$B = (X^T X)^{-1} \cdot X^T \cdot Y, \tag{1}$$

where B is a vector of the column values of the coefficients of multifactorial polynomial model; X is an experimental plan-matrix; X^T is a transposed of the experimental plan-matrix; Y is a column vector of the function response.

The verification of the adequacy of the obtained multi-factor regression levels is carried out by traditional methods of dispersion and correlation analysis. The adequacy of the Fisher criteria, used to compare the variances of two samples with normal distribution [16–18], was also verified.

Computer modeling enables visualization results based on computational experiment.

2 Constructing Mathematical, Experimental and Statistical Models to Predict the Lubricant Thickness

2.1 Experimental Research of the Lubricant Layer Thickness Under the Conditions of Local Contact

The main task of the experimental research is to determine the operating parameter of the friction pairs - rolling speed (condition of pure rolling) on the process of formation dynamics of the lubricating layer thickness in the central contact zone.

To carry out the experimental research, 6 marks of oils were used: (1) motor oil SAE 15w40 "LUX"; (2) motor oil M8G2K; (3) motor oil M10G2K; (4) oil I - 40; (5) universal oil motor - transmission EMT-8, (6) oil PROTEK EMT-8.

The speed changes ranged from 0 to 1.8 m/s; the volumetric temperature of the oil during the experiment ranged from 293.1 K to 343.1 K; the contact load ranged from 251.5 MPa to 399.4 MPa.

One of the main aspect of elastohydrodynamics (EHD) is point contact lubrication effect of rolling speed on the lubricant layer thickness. The change of the lubricant layer thickness, depending on the rolling speed, revealed the impact of rheological specifics of a wide range of lubricants on the productivity of roller bearings.

Clean rolling condition is necessary to avoid rising temperature of the lubricant and further reduces the viscosity at the inlet due to slipping, energy dissipation due to viscous friction [19]. To ensure a pure rolling (when the speed returns to its level) for the contact of the ball-disk, the correlation of the case V1 = V2 is reported in the expression:

$$V_1 - V_2 = \frac{2\pi}{60} \cdot (\omega_1 \cdot r_1 - \omega_2 \cdot r_2), \qquad (2)$$

where V_1, V_2 is a linear velocity of the ball and the disk; r_1, r_2 are the ball and the disk radiuses; ω_1 i ω_2 is an angular velocity of the ball and the disk.

In order to fulfill the condition of pure rolling, it is necessary to fulfill the condition of $\omega_1 = 3\omega_2$ and to ensure that the angular velocity of the ball exceeds the angular velocity of the disc thrice, then the expression 2 will take the form:

$$r_2 = 3r_1. \qquad (3)$$

Thus, the condition of pure rolling is provided by displacing the ball from the center of the disk rotation on the distance that is three times more than its own radius.

From theoretical research it is known the relationship between the coefficient of friction and the frequency of rotation of rolling elements.

56 experiments were conducted for each type of oil in two stages for different bulk temperatures. First, the value of the friction coefficient of the speed of rolling elements was determined. To determine the thickness of the lubricating layer for each type of oil, the value of the contact load in the friction pair, the bulk temperature of the lubricant material was also taken into account. As a result of experimental research, lubrication of oils of various composition and operational purpose was determined. The fragment of matrix input values and the quantities of the derived data of model experiment for oil PROTEK EMT-8 are shown in Tables 1 and 2 (N = 251.5 MPa; t = 293.1 K).

Table 1. Experimental and calculated values of coefficients of friction

№	Rotations, n, min^{-1}	Coefficient of friction F_T		Relative error calculations, %
		Experimental	Calculated	
1	5.4	0.0018	0.00179	0.55
2	6.5	0.0017	0.00166	2.35
3	7.5	0.0016	0.00157	1.88
4	7.8	0.0015	0.00153	2.0
5	8.6	0.0015	0.00149	0.66
6	8.8	0.0014	0.00144	2.86
7	9.8	0.0014	0.0014	0.0
8	10.5	0.0013	0.00134	3.08
9	11.2	0.0013	0.0013	0.0
10	12.4	0.0012	0.00121	0.83
...	...	...	...	...
54	68.2	0.0012	0.00115	4.17
55	70.3	0.0010	0.00099	1.0
56	78.0	0.0010	0.00095	5.0

The analysis of the experimental results showed that during the start-up phase in the friction pair, the kinetics of the formation of the lubricant layer thickness depends on the rolling bearings. With increasing rolling bearings, the lubricant layer thickness increases in the central contact zone. It leads to the establishment of appropriate lubrication modes (from the boundary to the hydrodynamic). The determining role is played by the kinematic viscosity of the lubricant, which basically depends on the base of the oils and the quantitative content of the additives [2].

The obtained experimental data can be used to create the experimental and statistical model that will enable a computational experiment. It will also make possible to determine the effect of the rheological features of a wide range of lubricants on the performance of roller bearings.

2.2   Experimental and Statistical Modeling Dependent on the Lubricant Layer Thickness from the Technical Parameters Specified in Local Contact

The results of experimental research determined that multiple factors X, affecting this process, are contact loads in paired friction (N), the frequency of rotation of rolling bodies (n, min^{-1}), the coefficient of friction (F_T) and the volume temperature of the lubricant (t, K). The initial parameter of the model (Y) is the value of the lubricant layer thickness (H_d).

Table 2. Experimental and calculated values of the lubricant layer thickness

№	Contact load, N, MPa	Rotations, n, min^{-1}	Coefficient of friction, F_T	Actual lubricant layer thickness, $H_d \cdot 10^{-6}$		Relative error calculations, %
				Experimental	Calculated	
1	251.5	7.8	0.0018	0.123	0.1231	0.08
2	251.5	9.8	0.0017	0.130	0.1298	0.15
3	251.5	11.2	0.0016	0.182	0.1817	0.03
4	251.5	13.4	0.0015	0.227	0.2269	0.01
5	251.5	16.4	0.0015	0.255	0.2548	0.02
6	251.5	20.4	0.0014	0.263	0.2635	0.05
7	251.5	24.8	0.0014	0.298	0.2981	0.01
8	251.5	28.4	0.0013	0.342	0.3423	0.03
9	251.5	34.2	0.0013	0.360	0.3596	0.04
10	251.5	38.8	0.0012	0.381	0.3808	0.02
...	...	...	...	...	...	...
54	399.4	54.5	0.0012	0.4423	0.4423	0.00
55	399.4	60.0	0.0010	0.4712	0.4712	0.00
56	399.4	67.5	0.0010	0.5769	0.5769	0.00

Modeling was carried out in two phases. At the first stage, the dependence of the coefficient of friction on the number of revolutions in the form of a single-factor regression model was shown as a dependence $F_T = f(n)$.

A general view of a single-factor model is:

$$F_T = \left(A_0 + \sum_{i=1}^{5} A_i \cdot n^i \right) \cdot 10^{-5}, \tag{4}$$

where A_i are coefficients of polynomial.

The coefficients of the model were calculated using Excel, as a result one-factor model was obtained:

$$F_T = (227.1 - 9.2982 \cdot n + 0.445071 \cdot n^2 - 0.0111312 \cdot n^3 + 0.000156978 \cdot n^4$$
$$- 0.000000532998 \cdot n^5) \cdot 10^{-5}. \tag{5}$$

This model enables to find the values of the coefficients of friction for an arbitrary value of the rotational speed of rolling bodies using oil of different operational purposes. The fragment of the experimental and calculated one-factor model of the coefficients of friction for PROTEK EMT-8 oil are shown in Table 1.

Figure 1 shows a graphical interpretation of the results of modeling the dependence of the coefficient of friction on the number of revolutions for oil PROTEK EMT-8.

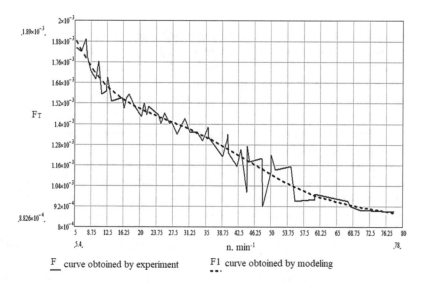

F   curve obtoined by experiment   F1   curve obtoined by modeling

Fig. 1. The comparison of field experiment and computational experiment results to determine the dependence of friction force on the number of rotations.

The verification of the adequacy of the obtained one-factor models was carried out by comparing the results of calculating the coefficients of friction obtained on the model with experimental data.

Fisher criterion value was determined for the number of degrees of freedom $d_{f_1} = d_{f_2} = 55$ and the level of significance $\alpha = 0$. The Fisher's criterion of table value, according to the data [18], equals $F_{kp} = 1.607289$ The calculated values $\left(F_{po3}\right)$ equal 1.0596. The condition was fulfilled $F_{po3} \leq F_{kp}$, then the hypothesis about the adequacy of the results obtained by modeling, the results obtained through field experiments, is confirmed by 5% of significance level.

The obtained functional dependence will be used as one of the factors of the multifactorial model. The general view of the multifactorial model is:

$$Y = \sum_{j=1}^{3} A_j \cdot x_j + \sum_{j=1}^{3} \sum_{i=1}^{3} B_{ij} \cdot x_i \cdot x_j + \sum_{k=1}^{3} C_k \cdot x_k^2. \tag{6}$$

where A_i, B_{ij}, C_k are polynomial coefficients.

Let us provide the following variables: $x_1 = N$ is a load at the point of contact; $x_2 = F_T(n)$ is a functional dependence of the coefficient of friction on the number of rotational speed; $x_3 = t$ is temperature; $Y = H_d$ is a true (actual) thickness, $\cdot 10^{-6}$.

Using a method of least squares a multi-factor regression model was obtained:

$$H_d(N, n, t) = 0.0012 \cdot N - 609.058 \cdot F_T(n) + 0.0667 \cdot t - 1.647 \cdot N \cdot F_T(n) + 0.00002 \cdot N \cdot t \\ - 0,3856 \cdot F_T(n) \cdot t + 0.00017 \cdot N^2 + 55911.25 \cdot F_T^2(n) - 0.00073 \cdot t^2. \tag{7}$$

This model enables to find the true lubricating layer thickness for arbitrary values of the parameters of the technological process using any oil. The fragment of the experimental and calculated by values model of the lubricant layer thickness for PROTEK EMT-8 oil is shown in Table 2.

Fisher's criterion of the tabular value, according to the data [18], equals $F_{kp} = 2.69$. Value calculation of the Fisher's criterion (F_{po3}) equals 2.28. The condition is fulfilled $F_{po3} \le F_{kp}$, then the hypothesis of the adequacy of the experimental results of the created model is confirmed.

To confirm the adequacy of the multi-factor experimental and statistical model, the multiplicity correlation coefficient was calculated. It characterizes the tightness of the connection between the experimental and calculated values of the lubricant layer thickness. The value of the correlation coefficient for PROTEK EMT-8 oil is equal 0.985. For all other types of oils, it varies from 0.9 to 0.985. It indicates a high level of connection between the experimental values of the output variable and the values obtained by the modeling system. The fragment of the values of the ration error of the lubricant layer thickness in the calculations on the models is shown in Table 2. It does not exceed 5% for all values.

Therefore, the resulting polynomial models adequately describe the dependence of the lubricant layer thickness on the load at the point of contact, the friction coefficient, which is a function of the number of revolutions, and the temperature for six types (kinds) of oils. The created experimental and statistical models can be used to make decisions and predict the properties of the lubricant from the specified technical parameters in the local connection.

2.3 Forecasting the Properties of the Lubricating Material from the Specified Technical Parameters in Conditions of Local Contact

Graphical analysis of regression equations is an efficient and universal tool for their visualization.

For graphical interpretation of technological solutions of mathematical models special computer programs were developed. For example, the "MathCad" program which enables to visualize the regression equations and represent the surface of the response in axonometry, or in projections on the reference frame. It is also possible to construct a projection of the contour lines of the surface response in the coordinates of the factors $X_i X_j$ indicating the pair combinations of independent variables of factors in the regression equation. To construct "quasi-factor" curves on a square in the model, only two of all factors and their combinations are left and other factors are sometimes fixed at some level.

Figure 2a shows the predicted values of the thickness of the lubricant layer for the developed mathematical model in the form of response of curves. Each curve in the square provides a certain value of the thickness of the lubricating layer which can be obtained by changing the above-mentioned factors of influence. The response surface on the cube of the actual thickness values, depending on the number of revolutions and temperature of the PROTEK OEM-8 oil, is shown in Fig. 2b. The type of surface indicates the stability of the investigated process, the absence of stochastic effects, and makes it possible to optimize the process parameters for each type of oil.

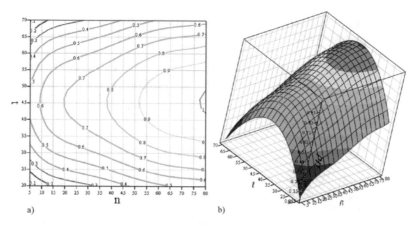

a)   b)

Fig. 2. Forecasted values of the thickness of the lubricant layer for the developed mathematical model in the form of: (a) response curves; (b) the response surface.

Graphical analysis of the obtained multicriteria experimental and statistical model shows that when the temperature of the lubricant material is t = 318.1 K, the process changing takes place, namely: up to t = 318.1 K the actual thickness increases, and after it begins to decrease. Increasing the number of revolutions leads to an increase in the actual thickness of lubricating layer in contact.

3   Conclusions

The experimental studies of the lubricant layer thickness under the conditions of local contact made it possible to determine the main factors that influence the rheological features of a wide range of lubricants on the suitability of roller bearings.

Such factors determine the pressure at the point of contact, the coefficient of friction, which, in its turn, depends on the number of revolutions and the temperature. The processing of the experimental results by the LSM method enabled to develop multivariate experimental and statistical regression models as the dependence of the actual thickness of the lubricating material on the determined factors. The adequacy of the models was checked by Fisher's criterion and correlation coefficients. Estimated values of Fisher's criteria for all models are less tabular, indicating that the obtained polynomials adequately describe the investigated process.

The obtained values of the correlation coefficients are from 0.9 to 0.985, which indicates a high level of connection between the experimental values of the output variable and the values obtained by simulating the thickness of the lubricant layer. The ration error of models does not exceed 5%.

Graphical interpretation of the results enabled to visualize the developed experimental and statistical models and to construct "quasi-factor" curves on the square and the surface of the response on the cube. Graphical analysis showed that when the temperature of the lubricating material reaches - 45 °C, the change in the thickness of the lubricating layer occurs, which initially increases at first, and at a temperature of 45 °C it begins to decrease. Increasing the number of revolutions leads to an increase in the actual thickness.

The created experimental and statistical models enable to predict the rheological properties of the lubricating material from the specified technical parameters under the conditions of local contact.

References

1. Mikosiyanchik, O.O.: Estimation of tribotechnical parameters of lubricating materials at the limit masking in conditions of local contact: dis. Cand. tech Sciences: 05.02.04; National trans un - K. (2006)
2. Savchuk, A.M.: Kinetics of change of lubricating properties of transmission and engine lubricants in conditions of abundant and limited lubrication: dis. Cand. tech Sciences: 05.02.04; National trans un - K. (2010)
3. Xiao, J., Wu, Y., Zhang, W., Chen, J., Zhang, C.: Friction of metal-matrix self-lubricating composites: relationships among lubricant content, lubricat-ing film coverage, and friction coefficient. Friction 1–14 (2019)
4. Altay, O., Gurgenc, T., Ulas, M., Özel, C.: Prediction of wear loss quantities of ferro-alloy coating using different machine learning algorithms. Friction 1–8 (2019)
5. Varenberg, M., Kligerman, Y., Halperin, G., Nakad, S., Kasem, H.: Assessing workability of greased bearings after long-term storage. Friction 1–8 (2018)
6. Malekan, A., Rouhani, S.: Model of contact friction based on extreme value statistics. Friction 1–13 (2018)
7. Evdokimov, A., Kursin, A.: Modeling of chemical and technological processes (experimental-statistical models): educational capability VSTEE SPbGUPTD, St. Petersburg (2018)
8. Voznesensky, V., Lyashenko, T.: Experimental-statistical modelling in computational materials science. Odessa: Astroprint (1998)
9. Voznesensky, V., Kovalchuk, A.: Decision making on statistical models. Statistics (1978)

10. Hartman, T., Klushin, D.: Fundamentals of computer simulation of chemical-technological processes textbook for high schools. ICC "Aka-demkniga", Moscow (2008)
11. Smirnov, V., Evstigneev, A.: Modeling efficiency in building material science. Int. Sci. J. **6**(48), 135–137 (2016). Part 2
12. Hartmann, T., Sovetin, F., Losev, V.: A Modern approach to modernization of chemical production on the basis of the application of models of modeling software chemistry, no. 12, pp. 8–10 (2008)
13. Bugaeva, L., Boyko, T., Beznosyk, Y.: System analysis of chemical and technological complexes. Interservice, Kyiv (2017)
14. Radchenko, S.: Mathematical modeling of technological processes in mechanical engineering. ZAO Ukrspetsmontazhproekt, Kyiv (1998)
15. Boyko, T., Abramova, A.: Mathematical modeling and optimization of processes of chemical technology, Kyiv (2016)
16. Vasilenko, O., Sencha, A.: Matematical-statistical methods of analysis in applied research: teaching. Manual - Odessa: ONAT them. O. S. Popova (2011)
17. Kublanov, M.: Test of the adequacy of mathematical models scientific novelty MITHA, no. 211, pp. 2–36 (2015)
18. Distribution tables. http://statsoft.ru/home/textbook/modules/sttable.html. Accessed 15 Apr 2019
19. Milanenko, O.A.: Lubricating the oils in the point contact of friction under conditions of heavy lubrication and lubricating fasting: dis. Cand. tech Sciences: 05.02.04. National Aviation University, Kyiv (2000)

Simulation of Working Process
of the Electronic Brake System
of the Heavy Vehicle

Dmytro Leontiev[1](✉) , Valerii Klimenko[1] ,
Mykola Mykhalevych[1] , Yevhen Don[1] , and Andrii Frolov[2]

[1] Kharkiv National Automobile and Highway University,
Department of Automobiles named after A.B. Gredeskul,
Yaroslava Mudrogo st, 25, Kharkiv 61002, Ukraine
dima.a3alij@gmail.com
[2] Kharkiv Scientific Research Institute Examinations named
Dist. Prof. N.S. Bokariusa, Zolochevskaya street, 8a, Kharkiv 61177, Ukraine

Abstract. This work analyzes and summarizes the results of the study, which show the level of influence of pressure change rate in an electro-pneumatic brake drive on the braking process of the wheels of a multi-axle vehicle. We have also performed mathematical modeling of the processes that take place in electro-pneumatic brake actuator during braking of a multi-axle vehicle affecting the dynamics of its movement. The purpose the study is to choose a rational rate of pressure change in an electro-pneumatic brake actuator of a multi-axle vehicle in the emergency braking mode. This work shows the effect of the mass of a multi-axle vehicle on deceleration value achieved in the process of braking. Here we also present the results of modeling the processes that take place in an electro-pneumatic brake drive of a multi-axle vehicle. This work also determines the degree of influence of the slope of the road surface on the intensity of braking of the wheels. Here we also describe the method for changing in rate of filling a brake chamber with a working fluid (air) upon modeling operation processes in an electro-pneumatic brake drive of a multi-axle vehicle.

Keywords: Electro-pneumatic braking system · Automated system ·
Vehicle braking dynamics · Heavy vehicle · Multi-axle vehicle

1 Introduction

There are many scientific publications devoted to research of the braking dynamics of multi-purpose vehicles equipped with pneumatic brake drive. These works show that the braking distance of vehicles of this class can be reduced to 20% due to reduction in the operating time of electro-pneumatic brake drives. At the same time, scientific and technical literature devoted to operation of automated braking control systems often focuses on the fact that such systems increase braking time of wheeled vehicles (WV). Therefore, there is an issue on selection of rational law of pressure change in electro-pneumatic devices of braking system used in heavy vehicles upon designing of automated braking systems.

2   Analysis of Publications

Several tasks should be solved upon designing a modern electro-pneumatic brake system (EBS) for a heavy vehicle:

- choice of criteria for triggering of EBS modulators;
- choice of the principle for controlling EBS modulators (considering the nature of the pressure change in the brake electro-pneumatic actuator).

The first task is solved by experimental study of the rolling process of the wheels of the vehicle [1–4]. In this case, the major issue is the choice of threshold values that ensure the quality of the automated brake control system. The second task relates to service life and cost of the automated brake control system. It is solved by achieving a rational correlation between the performance of the braking system and the complexity of implementing its control algorithms [5–8]. It should be noted that the algorithms for controlling electro-pneumatic pressure modulators play an important role in choosing control principle and are not widely covered in the scientific and technical literature. The analysis has showed that the choice of the law of pressure change in electro-pneumatic modulators of the brake system is very critical since it affects the overall braking efficiency of wheeled vehicle.

According to the analysis of scientific and technical literature [5–12], heavy vehicles use pressure modulator control that is based on three basic principles: IR (Individual regulation); MIR (Modified regulation); DIR (Diagonal regulation) IR or DIR principles ensure high braking performance. However, their application in guiding axles of heavy vehicle doesn't guarantee stable and controllable driving [10–12]. It should also be noted that these principles complicate the design of the brake system of wheeled vehicle due to increasing in number of control devices (pressure modulators).

MIR principles, as well as the DIR principles, ensures a high level of stability of wheeled vehicles by reducing the efficiency of the braking of the wheel transmission vehicle and is based on the low-threshold regulation of the change in angular rotation speed of the wheel. These principles ensure sufficient braking efficiency of guiding axle of wheeled vehicle. It should also be noted that MIR principle is not always based on individual pressure modulators. For example, the MIR principle can be based on brake drive with one electro-pneumatic pressure modulator as well as several electro-pneumatic devices on one axle of a wheeled vehicle. Analysis of control principles suggests that it is sufficient to use the MIR principle as a basic control principle to maintain the controllability and stability of a heavy vehicle with a large number of axles for further study of operation of electro-pneumatic braking system of heavy vehicle.

3   Study of Electro-Pneumatic Drive System of Heavy Vehicle

Simulating study of electro-pneumatic drive system of heavy vehicles depends on mathematic model of its movement dynamics. Adopted mathematical model should implement various laws of pressure change in electro-pneumatic pressure modulators of brake drive of heavy vehicle.

Based on the approach to the dynamics of wheeled vehicle motion proposed in work [11] and the approach to the interaction of automobile wheels of wheeled vehicle based on creep theory [12, 13], for the different loading conditions of the vehicle, we adopt the following mathematical model for determining the deceleration of heavy vehicle:

$$
j_a = \frac{g \cdot \left(D_0 \cdot \left(\sum_{i=1}^{n} A_{1\,i}^{left} \cdot y + \sum_{j=1}^{m} A_{2\,j}^{left} \cdot x \right) + D_1 \cdot \left(\sum_{i=1}^{n} A_{1\,i}^{right} \cdot y + \sum_{j=1}^{m} A_{2\,j}^{right} \cdot x \right) \right)}{2 \cdot B - h_g \cdot \left(2 \cdot B \cdot C_2 \cdot \left(\sum_{i=1}^{n} A_{1\,i}^{left} - \sum_{j=1}^{m} A_{2\,j}^{left} \right) + C_1 \cdot \left(\sum_{i=1}^{n} A_{1\,i}^{right} - \sum_{j=1}^{m} A_{2\,j}^{right} \right) \right)} , 0 \quad (1)
$$

where x and y are longitudinal coordinate positions of the center of gravity of heavy vehicle, m;

h_g – is vertical position of the center of gravity of a heavy vehicle, m;

B – is the averaged cross-sectional base of the wheeled vehicle, m;

D_0, D_1, C_1, C_2, C_3 and C_4 – are coefficients determined from the dependencies:

$$
D_0 = 2 \cdot B \cdot C_2 + \frac{C_3 \cdot C_4}{g}; \qquad D_1 = C_1 - \frac{C_3 \cdot C_4}{g};
$$

$$
C_1 = B \cdot \cos\beta - 2 \cdot h_g \cdot \sin\beta; \qquad C_2 = 1 - \frac{C_1}{2 \cdot B}; \qquad (2)
$$

$$
C_3 = 2 \cdot h_g \cdot \cos\beta + B \cdot \sin\beta; \qquad C_4 = V_x \cdot \omega_a - x \cdot \frac{d\omega_a}{dt}.
$$

where V_x – is the longitudinal velocity of the vehicle, m/s;

ω_a – is angular speed of the vehicle, c^{-1};

$\frac{d\omega_a}{dt}$ – is angular acceleration of a wheeled vehicle, c^{-2};

β – is angle of inclination of the road surface to the sidewalk, degrees;

$\sum_{i=1}^{n} A_{1\,i}$ and $\sum_{j=1}^{m} A_{2\,j}$ are parameters based on the realized couplings of the heavy

vehicle axles and determined from the dependencies (3)–(6), for the axles located in front and behind relative to the center of gravity of the vehicle, respectively.

$$
\sum_{i=1}^{n} A_{1\,i}^{left} = \sum_{i=1}^{n} \frac{f_{1\,i}^{left}}{\lambda_{1\,i}^{left}(L_a + x_i)}, \qquad (3)
$$

$$
\sum_{j=1}^{m} A_{2\,j}^{left} = \sum_{j=1}^{m} \frac{f_{2\,j}^{left}}{\lambda_{2\,j}^{left}(L_a + y_j)}, \qquad (4)
$$

$$
\sum_{i=1}^{n} A_{1\,i}^{right} = \sum_{i=1}^{n} \frac{f_{1\,i}^{right}}{\lambda_{1\,i}^{right}(L_a + x_i)}, \qquad (5)
$$

$$\sum_{j=1}^{m} A_{2j}^{right} = \sum_{j=1}^{m} \frac{f_{2j}^{right}}{\lambda_{2j}^{right}\left(L_a + y_j\right)},$$

(6)

where x_i and y_i are longitudinal coordinate positions of axles of the vehicle relative to the conditional longitudinal coordinates x and y of the center of gravity of heavy vehicle, m;

L_a – is conditional wheelbase of heavy vehicle ($L_a = x + y$), m;

λ_{1i}^{left}, λ_{1i}^{right}, λ_{2j}^{left} and λ_{2j}^{right} are the corresponding weight distribution factors between the front and rear left and right wheels of heavy vehicle.

$$\lambda_{1i}^{left} = \frac{\sum_{k=1}^{n} R_{z1k}^{left}}{R_{z1i}^{left}}, \qquad i = \overline{1, n},$$

(7)

$$\lambda_{1i}^{right} = \frac{\sum_{k=1}^{n} R_{z1k}^{right}}{R_{z1i}^{right}}, \qquad i = \overline{1, n},$$

(8)

$$\lambda_{2j}^{left} = \frac{\sum_{k=1}^{m} R_{z2k}^{left}}{R_{z2j}^{left}}, \qquad j = \overline{1, m},$$

(9)

$$\lambda_{2j}^{right} = \frac{\sum_{k=1}^{m} R_{z2k}^{right}}{R_{z2j}^{right}}, \qquad j = \overline{1, m},$$

(10)

where R_{z1i}^{left} and R_{z1i}^{right} are corresponding loads on the left and right axles of the heavy vehicle, located in front with respect to coordinate of its weight center, N;

R_{z2j}^{left} and R_{z2j}^{right} are corresponding loads on the left and right axles of heavy vehicle, located in the rear with respect to coordinate of its center of gravity, N;

It should be noted that the relevant loads are part of the left and right total loads on the front and rear axles of a heavy vehicle.

$f_{1i}^{left}, f_{1i}^{right}, f_{2j}^{left}$ and f_{2j}^{right} are corresponding realized couplings of the left and right wheels of heavy vehicle located in the front (i) and rear (j) with respect to coordinate of the center of gravity

$$f_{1i}^{left} = f\left(C_{x_i}^{left}, \xi_{x_i}^{left}, r_{d_i}^{left}, R_{z1i}^{left}\right),$$

(11)

$$f_{1i}^{right} = f\left(C_{x_i}^{right}, \xi_{x_i}^{right}, r_{d_i}^{right}, R_{z1i}^{right}\right),$$

(12)

$$f_{2j}^{left} = f\left(C_{x_j}^{left}, \xi_{x_j}^{left}, r_{d_j}^{left}, R_{z2j}^{left}\right),$$

(13)

$$f_{2j}^{right} = f\left(C_{x_j}^{right}, \xi_{x_j}^{right}, r_{d_j}^{right}, R_{z2j}^{right}\right), \tag{14}$$

where: $C_{x_i}^{left}$ and $C_{x_i}^{right}$ – are respective roll stiffness of right wheels of heavy vehicle that are located in the front relative to coordinates of its center of gravity, N m/radius;

$C_{x_j}^{left}$ and $C_{x_j}^{right}$ – are respective roll stiffness of right and left wheels of heavy vehicle located from behind relative to coordinates of its center of gravity, N m/radius;

$\xi_{x_i}^{left}$ and $\xi_{x_i}^{right}$ – are respective torsion angles of right and left wheels of heavy vehicle that are located in the front relative to coordinates of its center of gravity, radius;

$\xi_{x_j}^{left}$ and $\xi_{x_j}^{right}$ – are respective torsion angles of right and left wheels of heavy vehicle that are located from behind relative to coordinates of its center of gravity, radius;

$r_{d_i}^{left}, r_{d_i}^{right}, r_{d_j}^{left}$ and $r_{d_j}^{right}$ – are respective dynamic radiuses of right and left wheels located in the front (i) and from behind (j) relative to coordinates of center of gravity of vehicle, m;

Based on accepted mathematical model of motion dynamics of a heavy vehicle that considers the level of loading of a vehicle, it is easy to simulate the nature of the motion of such a vehicle, taking into account the slope of the road surface towards roadside, by determining of torsion angles of tires of heavy vehicle through the equation of motion dynamics of wheels. These angles depend on the nature of the pressure change in the brake cylinders of vehicle brake drive.

To solve the gas dynamics equations, we adopt the classical methods described in works [8, 11]. These methods represent pressure change during the working process of the brake drive as increase, holding and emptying phase of its throttling links, including cyclic mode. General view of the equations that describe the dynamics of pressure variation in throttling links, for example, front independent circuit with one axial pressure modulator (Fig. 1), is as follows.

Figure 1 contains the following abbreviations: V_{k11}, V_{k12} are the capacities of the brake chambers of one axle of heavy vehicle, m^3; P_{k11}, P_{k12} are pressure values in the brake chambers of the circuit, Pa; r_{k11}, r_{k12} are cavities throttling the flow of the working medium in the brake chambers of the circuit; Y_{k11}, Y_{k12} are corresponding throttling links (brake chambers) of circuit; G_6, G_7 are instantaneous mass flows of air in the brake chambers of the circuit; V_{m1}, V_{m11}, V_{m12} are the volumes of cavities of connecting pipelines of the circuit, m^3; P_{m1}, P_{m11}, P_{m12} are pressures in the connecting pipelines of the circuit, Pa; r_{m1}, r_{m11}, r_{m12} are respective cavities that throttle the flow of the working medium in the connecting pipelines of circuit; Y_{m1}, Y_{m11}, Y_{m12} are corresponding throttling links (connecting pipeline) of circuit; G_2, G_4, G_5 are corresponding instantaneous mass flow rates of air in the connecting pipelines of the circuit; V_{M11} is a volume of the cavity of the axial modulator, m^3; P_{M11} is pressure in the axial pressure modulator, Pa; r_{M11} is cavity that throttles the flow of working medium into a pressure modulator; Y_{M11} is corresponding throttling link (pressure modulator) of circuit; G_3 is instantaneous mass flow of air in the pressure modulator; V_{V1} is a volume of the cavity of the air tank, m^3; P_{V1} is pressure in the air tank, Pa; r_{V1} is cavity that

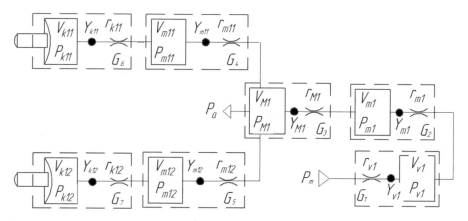

Fig. 1. Circuit of the one of front circuits of electro-pneumatic brake drive with axial pressure modulators

throttles the flow of working medium into a air tank; Y_{V1} is corresponding throttling link (air tank) of circuit; G_1 is instantaneous mass flow of air in the air tank.

$$
\begin{cases}
\frac{dP_{V1}}{dt} = \frac{(G_1-G_2)\cdot k\cdot R\cdot T}{V_{V1}} \\
\frac{dP_{m1}}{dt} = \frac{(G_2-G_3)\cdot k\cdot R\cdot T}{V_{m1}} \\
\frac{dP_{M1}}{dt} = \frac{x_{M1}\cdot(G_3-G_4-G_5)\cdot k\cdot R\cdot T}{V_{M1}} \\
\frac{dP_{m11}}{dt} = \frac{x_{M1}\cdot(G_4-G_6)\cdot k\cdot R\cdot T}{V_{m11}} \\
\frac{dP_{m12}}{dt} = \frac{x_{M1}\cdot(G_5-G_7)\cdot k\cdot R\cdot T}{V_{m12}} \\
\frac{dP_{k11}}{dt} = \frac{x_{M1}\cdot G_6\cdot k\cdot R\cdot T}{V_{k11}} \\
\frac{dP_{k12}}{dt} = \frac{x_{M1}\cdot G_7\cdot k\cdot R\cdot T}{V_{k12}}
\end{cases}
\tag{15}
$$

In Eq. (15), parameter x_{M1} characterizes operation mode of modulator of electro-pneumatic brake drive. Depending on regulation phase (filling/holding/emptying), the signal may have the following values: $x_{M1} = 1$ upon filling of throttling link, $x_{M1} = 0$ upon holding pressure in throttling link and $x_{M1} = -1$ upon emptying of throttling link. When $x_{M1} = 0$, the pressure in modulator should be assumed equal to the pressure at previous calculation stage.

Constant coefficients k, R, T in the Eq. (15) are taken according to the scientific and technical literature [8, 11] (adiabatic index $k = 1.4$; gas constant $R = 287.14$ m^2/(c$^2\cdot K$); the absolute temperature of the working medium before throttle $T = 283\ K$).

Adopted mathematical model of the dynamics of pressure variation in electro-pneumatic brake drive can be easily executed by means of software complex MATLAB package SIMULINK (Fig. 2). Output data for the simulation of the electro-pneumatic brake drive are taken in accordance with specifications of mobile laboratory [14] of A. B. Gredeskul Department of Vehicles where the individual pressure modulators were installed.

The results of the simulation of the pressure dynamics in one of the front (Fig. 3) and one of the rear (Fig. 4) of circuits of electro-pneumatic brake drive in ABS mode have showed that the decrease in the speed of wheeled vehicle results in decrease of intensity of emission of the working medium from the electro-pneumatic brake drive and frequency of its operation increases due to increase in the value of adhesion of tire with road surface that has direct effect on vehicle control (j_a).

The analysis of the simulation of operation process in the front and rear circuits of the electro-pneumatic drive of the brake system of a heavy vehicle suggests that pressure drop is more considerable in drive at the beginning of braking process in comparison with pressure drop at the end of braking process due to dynamic processes during adhesion of wheel tire to road surface.

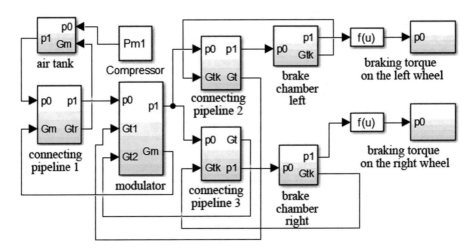

Fig. 2. Structural and logical scheme of implementation of the front/rear circuit of the electro-pneumatic brake drive

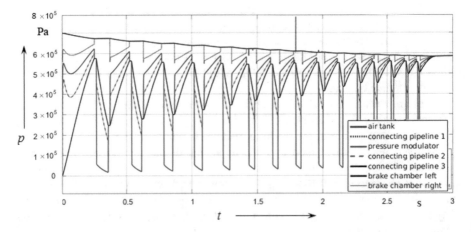

Fig. 3. Dynamic processes of pressure variation in one of front circuits of electro-pneumatic brake drive of heavy 1 vehicle during the operation of the automated ABS system (angle of inclination of road surface towards roadside is 1°.)

The analysis of the intensity of filling of circuits of electro-pneumatic, as shown on Figs. 3 and 4, suggests that the rapid filling of throttling links of the drive does not lead to a significant increase in braking efficiency of heavy vehicle. This fact is confirmed by results of the research published in the generally available scientific and technical literature [4–8] since automated brake control systems reduce overall braking performance of the wheeled vehicle due to wheel brake release time.

Experimental studies shown in Fig. 8 suggest the reduction in wheel brake release time prevents quick unblocking of the wheel. Therefore, it is necessary to use other approaches to the organization of algorithms for the operation of automated systems that would provide efficient braking of vehicles equipped with electro-pneumatic brake drive.

One of these approaches is to extend the processes of filling of throttling links of the brake system drive within the range from 0.2 to 0.5 MPa, since such extension is limited to 0.6 s. by nugatory documents and standards for a pneumatic drive at the beginning of the braking process. The simulation of this approach showed (Figs. 5 and 6) that the number of cycles of an automated brake control system decreases by an average of 25% without significantly changing the braking performance of a heavy vehicle (the braking performance decreases by no more than 1%).

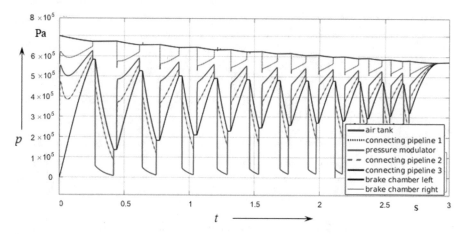

Fig. 4. Dynamic processes of pressure variation in one of rear circuits of electro-pneumatic brake drive of heavy 1 vehicle during the operation of the automated ABS system (angle of inclination of road surface towards roadside is 1°.)

The analysis of the simulation of vehicle movement dynamics (Fig. 7), considering processes that take place in brake drive suggest that the decrease of average pressure in circuits connected to the rear axles of heavy vehicle is conditioned by changes in the physical properties during adhesion of tires to road surface. The simulation of the vehicle movement dynamics also showed that the extension of the processes of filling the brake chambers during the operation of the electro-pneumatic brake drive (Figs. 5 and 6) reduces the load on the driver's vestibular apparatus and creates more comfortable conditions in case of emergency braking of the vehicle (Fig. 7).

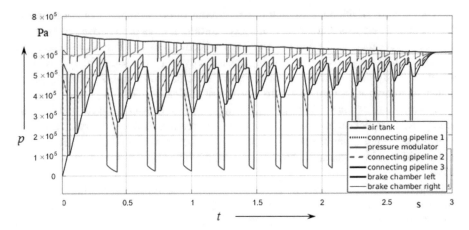

Fig. 5. Dynamic processes of pressure variation in one of the front circuits of electro-pneumatic brake drive of heavy vehicle during the simulation of the nature of the stepped pressure change in the brake drive (angle of inclination of the road surface to roadside is 1°)

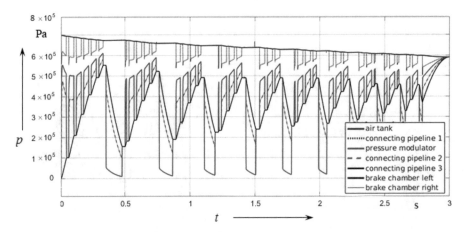

Fig. 6. Dynamic processes of pressure change in one of the rear circuits of electro-pneumatic brake drive of heavy vehicle during simulation of the nature of the stepped pressure change in the brake drive (angle of inclination of the road surface to roadside is 1°)

During the simulation study, it was also found that the braking performance of the wheeled vehicle depends on overlapping of operation phases of the automated brake control system located in the rear and front circuits of the vehicle braking system (Fig. 7). During the simultaneous release of several circuits, the overall efficiency of vehicle's braking is reduced due to the simultaneous release of wheels brake.

The study of braking dynamics of vehicle in loaded condition showed that the decrease in the weight of the vehicle (Fig. 7) compared to maximum loading (Fig. 6) results in the increase of overall braking performance up to 20%. It should be noted the value of the average pressure in the brake drive decreases with the increase of

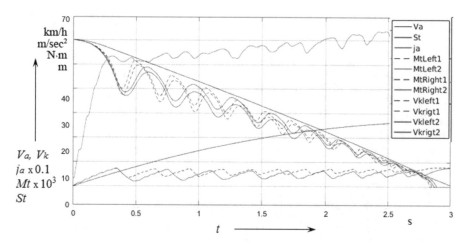

Fig. 7. Braking dynamics of heavy vehicle (angle of inclination of the road surface towards roadside is 1°, two front and two rear circuits of a four-wheeled vehicle are operated simultaneously)

deceleration (Figs. 2 and 7). This is due to the physical processes that occur in the area where tire adheres to road surface, which are described in the scientific and technical literature [8, 11–13, 15–18].

4 Experimental Studies of Operation Process of an Electro-Pneumatic Brake Drive During an Emergency Braking of Heavy Vehicle

The studies of operation process that occurs in the electro-pneumatic brake system of heavy wheeled vehicle during emergency baking (Fig. 8) confirmed theoretical studies shown in Figs. 5, 6 and 7.

It should be noted that during the experimental study (Fig. 8) the average pressure in drive was 0.45 MPa with a maximum permissible load of 25500 N on the automobile wheel due to operation of the automated system. It was not allowed to reduce the pressure in the system below 0.2 MPa during operation of the automated system. The upper limit of pressure was not limited and maintained at a level not lower than 0.6 MPa.

Experimental studies of wheel braking process under the influence of the electro-pneumatic brake drive during extension of brake chamber filling (Fig. 8) showed that such approach leads to a significant reduction in the load on the brake mechanism due to the decrease in the frequency of alternating loads upon application of braking force between tire and road surface.

In this case, larger pressure is observed in the brake chamber in comparison with that observed during braking under similar conditions without extension of filling of pneumatic brake chambers of the electro-pneumatic brake drive of heavy vehicle, depending on the value of tire slip relative to road surface due to the implementation of the braking force.

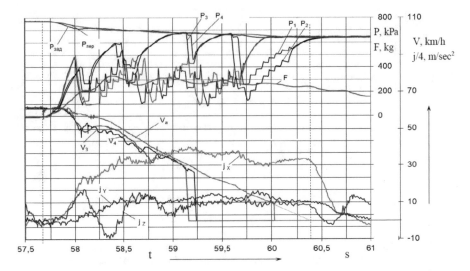

Fig. 8. Experimental highway studies of the operation of electro-pneumatic brake drive in ABS mode by extension of the brake chamber filling process during execution of the ABS algorithm

5   Conclusions

The simulation of the nature of the pressure change in the electro-pneumatic brake actuator of the primary braking system suggests that the most rational nature of the pressure change in throttling links in the emergency braking mode of wheeled vehicle is the rapid increase in pressure at the beginning of operation process (up to 0.3 MPa for a time of not more than 0.3 s) and a smooth increase in pressure during subsequent braking (the pressure should increase by 0.1 MPa within time equal to or not less than 1 s).

The intense pressure increase in throttling link leads to frequent tripping of the automated brake control system. Therefore, it is necessary to reduce the number of cycles of its operation and fairly high braking efficiency of heavy vehicle during designing an automated brake control system.

References

1. Mihalevich, N.G.: Sovershenstvovanie elektropnevmaticheskih apparatov tormoz-nogo privoda avtotransportnyih sredstv [Improvement of electropneumatic devices of a brake drive of vehicles]. Dissertation, Kharkiv National Automobile and Highway University (2009)
2. Popov, A.I., Nuzhnyiy, V.V.: Otsenka harakteristik elektro-pnevmaticheskogo tormoznogo privoda [Evaluation of the characteristics of an electropneumatic brake actuator]. Puti sovershenstvovaniya avtomobilya i ego apparatov/ Mosk. avtomob. – dor. in-t, pp. 35–40 (1988)

3. Nuzhnyiy, V.V.: Razrabotka elektropnevma-ticheskogo tormoznogo privoda avtotrans-portnogo sredstva [Development of an electropneumatic motor vehicle brake drive]. Dissertation, Donetsk National University (1996)
4. Krasyuk, A.N.: Sovershenstvovanie elektro-pnevmaticheskih sistem avtotransportnyih sredstv [Improving the electropneumatic systems of vehicles]. Dissertation, Kharkiv National Automobile and Highway University (2011)
5. Gurevich, L.V., Melamud, R.A.: Pnevmaticheskiy tormoznoy privod avtotransportnyih sredstv [Pneumatic brake drive for vehicles]. Transport, Moscow (1988)
6. Pchelin, I.K.: Dinamika protsessa tormozheniya avtomobilya [The dynamics of the process of braking the car]. Dissertation, Moscow Automobile and Road Construction State Technical University (1984)
7. Revin, A.A.: Teoriya ekspluatatsionnyih svoystv avtomobiley i avtopoezdov s ABS v rezhime tormozheniya [Theory of operational properties of cars and trucks with ABS in braking mode]. VolgGTU, Volgograd (2002)
8. Leontiev, D.: Systemnyi pidkhid do stvorennia avtomatyzovanoho halmivnoho keruvannia transportnykh zasobiv katehorii M_3 ta N_3 [System approach to the creation of automated brake control of vehicles of categories M_3 and N_3]. Dissertation, Kharkiv National Automobile and Highway University (2011)
9. Electronic braking system for Trailers with roll over protection. KNORR-BREMSE System for commercial Vehicles (Electronic resource) Product information. Site access mode. http://en.knorr-bremsesfn.com/systems/
10. Electronically Controlled Braking Systems for Trailers (Electronic resource): Pamphlet. A Division WABCO Standart GmbH. EBS (EPB) - 2013 edition 28p. - Electronic text data. WABCO 2015. 1 electronic opt. disk (DVD-ROM)
11. Turenko, A.N., Klimenko, V.I., Ryizhih, L.A., Bogomolov, V.A., Leontiev, D.N., Krasuk, A.N., Myhalevich, N.G.: Realizatsiya intellektualnyih funktsiy v elektronno-pnevmaticheskom tormoznom upravlenii transportnyih sredstv [The implementation of intellectual functions in the electron-pneumatic braking control of vehicles]. Kharkiv National Automobile and Highway University, Kharkiv (2015)
12. Levin, M.A., Fufaev, N.A.: Teoriya kacheniya defor-mirovannogo kolesa [The theory of rolling deformed wheels]. Nauka, Moscow (1989)
13. Leontiev, D., Ryizhih, L., Byikadorov, A.: Opredelenie prodolnoy realizuemoy silyi stsepleniya avtomobilnogo kolesa s opornoy poverhnostyu po krutilnoy deformatsii shinyi i ee zhestkosti [Determination of the longitudinal realizable force of adhesion of an automobile wheel with a bearing surface by torsional deformation of the tire and its rigidity]. Avtomobilnaya promyishlennost **10**, 20–25 (2014)
14. OAO «Minskiy avtomobilnyiy zavod» (2012) Avtorbus MAZ-256. Rukovodstvo po ekspluatatsii 256-0000020 RE [The bus MAZ-256. Operating Instructions 256-0000020 OI.]. Minskiy avtomobilnyiy zavod, Minsk (2012)
15. Denny, M.: The dynamics of antilock brake systems. Eur. J. Phys. **26**(6), 1007–1016 (2005)
16. Ersal, T., Fathy, H., Stein, J.: Structural simplification of modular bond-graph models based on junction inactivity. Simul. Model. Pract. Theory **17**, 175–196 (2009)
17. Oniz, Y., Kayacan, E., Kaynak, O.: A dynamic method to forecast the wheel slip for antilock braking system and its experimental evaluation. IEEE Trans. Syst. Man Cybern. Part B Cybern. **39**(2), 551–560 (2009)
18. Zheng, T.: Research on road identification method in anti-lock braking system. Procedia Eng. **15**, 194–198 (2011)

Defining Pre-emergency and Emergency Modes of Semiconductor Converters

L. Laikova$^{(\boxtimes)}$ ⓘ, T. Tereshchenko ⓘ, and Y. Yamnenko ⓘ

National Technical University of Ukraine "Igor Sikorsky Kyiv
Polytechnic Institute", Peremohy Av., 37, Kyiv 03056, Ukraine
laikova@ukr.net

Abstract. The paper is devoted to the problem of identifying malfunctions of semiconductor converter elements on the example of DC-DC converters with quasi-impedance link The definition of emergency modes (such as breakages, breakdowns and short circuits of passive elements of the converter) and pre-emergency modes (such as malfunctions caused by the gradual exit of the parameters of the elements beyond the permissible limits) is considered. It is proposed to perform the identification of the first type of malfunction according to the criteria of the mean square error and Euclidean distance for different states - the serviceable, which corresponds to the nominal values of the parameters of the elements, and a number of emergency ones. The identification of the second type of malfunctions is proposed to be performed at the average of the rectified voltage in the steady state and the time of the transient process. The questions of determination of the area of trouble-free operation are considered at the simultaneous change of several parameters of the scheme.

Keywords: Malfunctions Identifications · Diagnostics of Converters ·
DC-DC Converter with Quasi-impedance Link

1 Introduction

The state of the semiconductor converter is influenced by such factors as ambient temperature, operating frequency, operating currents and voltages, operating conditions, changes in the internal parameters of the elements, which sometimes leads to emergency or pre-emergency operation modes. One of the methods for obtaining diagnostic information about the operation mode of a semiconductor converter is the simulation of processes in it and the allocation of certain diagnostic indicators, which will become the basis for identifying its condition.

As the research shows, the change in the forms of currents and voltages time charts is evidenced by the appearance of a malfunction or a change in the parameters of the elements of a semiconductor converter [11, 12]. Therefore, the time dependencies of currents and voltage of converters should be used as diagnostic information.

The task of detecting emergency modes of the converter is successfully solved in many ways [13], but the detection of a pre-emergency mode due to the exit of converters elements parameters beyond the permissible limit is a much more complicated

© Springer Nature Switzerland AG 2020
A. Palagin et al. (Eds.): MODS 2019, AISC 1019, pp. 62–70, 2020.
https://doi.org/10.1007/978-3-030-25741-5_7

task, which involves a deep analysis of the circuit of the investigated converter in different operating modes and at different values of parameters.

The aim of this paper is to investigate the effects of a *DC-DC* converter with a quasi-impedance link in the form of output voltage charts in the time and spectral Walsh transformation areas, and to develop a recommendation on how to identify the current malfunction, as well as to research and define malfunctions caused by deviations of parameters of elements within the permissible limits, for timely detection of pre-emergency modes of the converter operation.

2   Converter Circuit

Converters with a quasi-impedance (QI) link are characterized by a wide range of regulation of the input voltage, the absence of jump of input current, high noise immunity. These characteristics caused their application in systems with alternative energy sources for converting the output voltages of these sources (usually constant and varying widely) into constant voltages with given parameters [1–4].

Figure 1 shows the scheme of the converter with a quasi-impedance (QI) link.

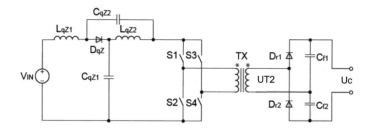

Fig. 1. The scheme of a quasi-impedance *DC-DC* converter

The converter contains a voltage source (V_{IN}), a bridge galvanically isolated *DC-DC* converter, the input of which is a quasi-impedance link (*qZS*) of two capacitors C_{qZ1} and C_{qZ2}, two inductors L_{qZ1} and L_{qZ2} and a diode D_{qZ}.

The *DC-DC* converter includes an inverter built on four MON-transistors (S1-S4 keys in Fig. 1), an isolation transformer TX, and a voltage doubler from two diodes D_{r1} and D_{r2} and two capacitors C_{f1} and C_{f2}.

3   Defining of Emergency Modes

The model of the quasi-impedance *DC-DC* converter is implemented in the branching package mathematical system *MATLAB R2014a - Simulink R2014a* (Fig. 2). The model contains a constant voltage source V_{IN} (*DC Voltage Source*), which is connected to the inverter bridge *Universal Bridge*. Between the source and the inverter passive *qZS*-link is present. Transistor control is carried out using the Control system control unit, which specifies the pulse control amplitude (units), their period (0.02 s), width (%) and phase

delay (sec). The bridge inverter is connected to the voltage doubler via a linear transformer T (*Linear Transformer*) with a transform coefficient of 1:4. As diagnostic
indicator for the circuit the diagram of the voltage of the secondary winding of the
transformer – *UT2* is selected.

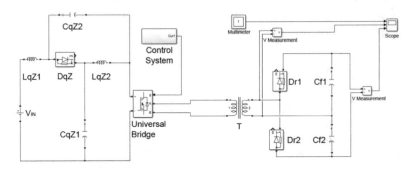

Fig. 2. Model of quasi-impedance DC-DC converter

The correct operation of the converter is obtained with the following parameters of
the model $V_{IN} = 15$ V; $L_{qZ1} = L_{qZ2} = 0.5$ μH, $C_{qZ1} = C_{qZ2} = 26.4$ μF; $C_{f1} = C_{f2} = 2.2$
μF (Fig. 3).

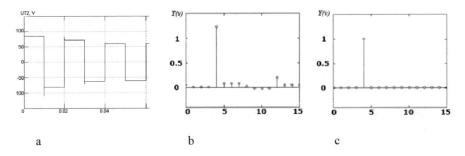

Fig. 3. Time dependences of the converter voltage at the nominal values of the parameters of
elements (a) and its spectra in transition (b) and steady state (c)

Since in the steady state the voltage *UT2* coincides with the Walsh function *wal*
(a, t), the spectrum of such a function will contain only 1 count at $v = a$ [5, 6]. This
suggests that determining the emergency state by comparing a known spectrum of the
normal state and the current one will take less time than comparing the time diagrams.
Walsh spectrum will be calculate for normalized to the steady-state value in the range of
two periods of change of voltage (0–0.04 s). We take this interval equal to $N = 16$
counts. Then the discrete time original function for Fig. 3a will be $y(x) = (1.58; 1.58;
1.54; 1.54; -1.96; -1.34; -1.34; -1.34; 1; 1; 1; 1; -1; -1; -1; -1)$. Walsh's spectrum $Y(v)$
is found as

$$Y(v) = \frac{1}{N} F_{wal} y(x), \tag{1}$$

where $F_{wal} = \begin{bmatrix} 1 & 1 \\ 1 & -1 \end{bmatrix}^{[4]}$ – the basic Walsh functions in the 4th degree of the Kronecker product [7–9] will contain 16 different values. In the analysis of the steady state, the Walsh spectrum will contain only one count [14].

In the diagnosis and analysis, the charts of the converter operation were analyzed at various possible emergency conditions for it, such as breakages, breakdowns, short circuits, and their Walsh spectra, calculated by the Eq. (1).

For all 10 types of malfunctions the measures of proximity of emergency and correct condition are calculated, Table 1.

Table 1. Proximity measures

Malfunction type	norms	1	2	3	4	5	6	7	8	9	10
Euclidean distance in time domain	0	4.814	3.126	0.571	5.032	0.891	2.674	0.481	0.452	14.955	9.901
Euclidean distance in the spectral domain	0	1.203	0.781	0.143	1.258	0.223	0.669	0.12	0.113	3.739	2.475
Mean-square error in time domain	0	1.448	0.611	0.02	1.582	0.05	0.447	0.014	0.013	13.978	6.126
Median error in the spectral domain	0	0.091	0.038	0.0013	0.099	0.003	0.028	0.001	0.0008	0.874	0.383

Analysis of Table 1 showed that the identification of the type of malfunction in the converter should be performed using the criterion of the Euclidean distance in the spectral region, since in this case it is obtained in 0.619/0.064 = 9.6 times greater relative difference, even for similar malfunction, compared with the case of analysis in time zone [14].

4 Defining of Pre-emergency Modes

In Tables 2, 3 and 4 is shown calculated according to the charts the average values of the rectified stress voltage in the steady state and the transition time at different inductance parameters the inductance L_{qZ} (Table 2), the capacitance C_f (Table 3), and the resistance R_{on} of the transistors (Table 4).

Table 2. Simulation results when changing capacitance $Lqz1$ ($Lqz2$)

Deviations L_{qZ} (in % of nom.)	−75	−50	−10	0	+10	+50	+75
U_c, V	196.9	191.7	204.7	205.4	205.4	200.8	200.3
t_{tp}, ms	17	24	10.5	10.5	13.5	30.15	30.17

Table 3. Simulation results when changing capacitance C_{f1} (C_{f2})

Deviations C_f (in % of nom.)	−75	−50	−10	0	+10	+50	+75
U_c, V	198.6	197.9	204.7	205.4	205.9	206.9	207
t_{tp}, ms	11.3	10.35	10.3	10.5	10.4	10.3	10.3

Table 4. Modeling results when changing resistance R_{on}

Deviations R_{on} (in % of nom.)	−75	−50	−10	0	+10	+50	+75
U_c, V	214.5	211.4	206.6	205.4	204.2	199.8	197
t_{tp}, ms	10.3	10.3	10.3	10.5	10.4	10.4	10.4

5   Analysis of Results

In Fig. 4 shows the voltage curves U_c and time t_{tp} when changing the parameters of the converter L_{qZ1} (L_{qZ2}), C_{f1} (C_{f2}), R_{on} according to Tables 3, 4 and marked points corresponding to the limit of trouble-free operation (in this case ± 10% of the nominal value is taken). As can be seen from Fig. 4, these two parameters uniquely characterize the type of malfunction.

In Table 5 shows the graphs of the dependence of the mean value of the rectified voltage and the time of the transient process on the value of the parameters L_{qZ1} (L_{qZ2}), C_{f1} (C_{f2}), R_{on}.

Assume that at the same time only one of the three parameters of the converter circuit changes. Then, based on the results of the simulation and the measured voltage values U_c and time t_{tp}, you can find the current value of this parameter and determine if it is approaching the given allowable limit.

For example, if only the inductance changes: measuring results follows: U_c = 200 V, t_{tp} = 30 ms. According to the charts of Table 5 (first line), the value of L_{qZ} = 0.8 µH is determined. If the permissible deviation limit is 10% of the nominal value, that is, the critical value of the inductance is equal to L_{qZ} = 0.55 µH, then the calculated value of 0.8 µH indicates a choke malfunction.

Note that measurement of one indicator (voltage U_c or time t_{tp}) is insufficient. In the example given, the measured value U_c = 200 V corresponds to two points on the chart of Table 5 (first line) - L_{qZ} = 0.4 µH and L_{qZ} = 0.8 µH, and only the value of t_{tp} = 30 ms can clearly determine that L_{qZ} = 0.8 µH.

A more complex option is the simultaneous change of several parameters, such as L_{qZ1} (L_{qZ2}), C_{f1} (C_{f2}), Ron within + 10%.

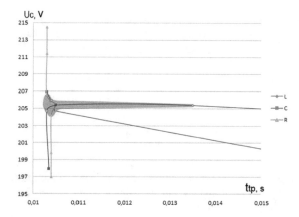

Fig. 4. Change the voltage *Uc* and time ttp when changing the parameters of the converter (Shaded area corresponds to the area of trouble-free operation with the simultaneous change of parameters by ± 10%)

Table 5. Modeling results

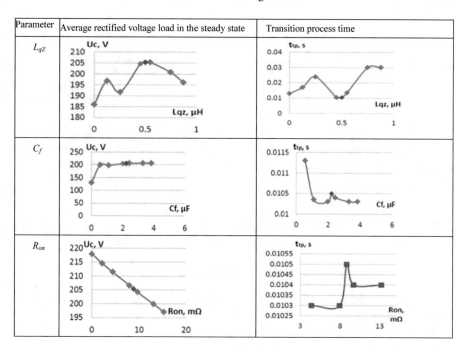

Parameter	Average rectified voltage load in the steady state	Transition process time
L_{qZ}		
C_f		
R_{on}		

To describe the function of three changes, it is advisable to apply the Brandon method [15], based on the processing of experimental data. According to the Brandon algorithm, the function of several variables is defined as:

$$\hat{y}(x_1, x_2, \ldots, x_n) = b_0 \prod_{i=1}^{n} f_i(x_i), \tag{2}$$

$$b_0 = \frac{1}{N} \sum_{u}^{N} y_{nu}, \tag{3}$$

where $f_i(x_i)$ – empirical regression functions, N – volume of sampling.

The value of b_0 is calculated as the average voltage U_c in the range of + 10% change of each parameter (see Table 4):

$$b_0 = \frac{1}{27} \sum_{j=1}^{27} y_j = 205, 11.$$

In Table 6 is marked x_1 - resistance R_{on} MOSFET, x_2 - capacitance C_{f1}, x_3 - inductance L_{qZ1}.

Table 6. Experimental values of Uc (B) for different combinations of parameters

x_3	0.45 µH			0.5 µH			0.55 µH		
x_2 x_1	1.98 µF	2.2 µF	2.42 µF	1.98 µF	2.2 µF	2.42 µF	1.98 µF	2.2 µF	2.42 µF
7.92 mΩ	205	206	206.4	205.9	206.56	207	205.9	206.75	207
8.8 mΩ	204.1	204.75	205.2	204.8	205.4	205.8	204.8	205.4	205.8
9.68 mΩ	203.1	203.57	204	203.7	204.24	204.6	203.7	204.26	204.65

Experimental data show that the dependence of U_c on the value of the parameters x_1-x_3, in the region + 10%, is linear and can be described by the equation of type:

$$U_c(x_i) = ax_i + b.$$

Using the method of least squares we can find coefficients a and b for each of the parameters [16].

For resistance R_{on}: a = –0.00625 and b = 1.056. The error is 0.0005 for the resistance values 7.92 mΩ and 0.0001 for 9.68 mΩ.

For a capacity C_{f1}: a = 0,0114 and b = 0,976. Error within the studied data has not been established.

For inductance L_{qZ1}: a = 0.03 and b = 0.985. The error is relevant for the values of the inductance of 0.45 µH, 0.5 µH and 0.55 µH. It is 0.0005, 0.001 and 0.0005, respectively.

According to Brandon's method, the dependence of the output voltage on such parameters as resistance R_{on}, inductance L_{qZ1} and capacity C_{f1}, is mathematically written as follows:

$$U_{\tilde{N}} = 205,11 \cdot (1,056 - 0,00625R_{on})$$
$$\cdot (0.0114C_{f1} + 0,976) \cdot (0,03L_{qz1} + 0,985) \,. \tag{4}$$

The maximum error of the Uc determination is 0.35%.

By Eq. (4), you can calculate the output voltage DC-DC converter without the involvement of simulation programs.

Defining a similar dependence on the time of the transient process is a much more complicated task due to its nonlinear dependence on the parameters of the elements of the scheme. For approximate estimation, it is proposed to determine the time of the transient process for each combination of the parameters of the scheme (Table 4) as the maximum of the set of received.

On the plane U_c i t_{tp} (Fig. 3), the calculated values determine the area of trouble-free operation while simultaneously changing the parameters by + 10%.

6 Conclusions

By Walsh's spectrum it is easy to determine the presence of emergency conditions in DC-DC converters - the spectrum for a correct condition has a unique nonzero Walsh spectrum count. The appearance of other nonzero components in the spectrum indicates a malfunction.

The proposed method for the determination of pre-emergency conditions involves assessing the state of the converter in two parameters - the average value of the rectified voltage load in the steady and the time of the transition process at various parameters of the elements. The assumption that at the same time only one of the parameters of the scheme changes, allows to determine the current deviation of the parameter from the nominal value by the values of the two diagnostic parameters indicated above. For the case of the simultaneous change of several parameters formed area of trouble-free operation, the output beyond which indicates the occurrence of malfunction.

References

1. Liivik, L.: Semiconductor Power Loss Reduction and Efficiency Improvement Techniques for the Galvanically Isolated Quasi-Z-Source DC-DC Converters (2015). https://digi.lib.ttu.ee/i/?2519
2. Gusev, O.O.: Viznachennya parametriv regulyatora v sistemi keruvannya DC/DC peretvoryuvachem z kvazi-impedansnoyu lankoyu za umovi stijkosti dlya malogo signalu [Determining of the controller parameters of the qZS DC/DC converter control system providing small signal]. Tech. Electrodyn. **5**, 18–23 (2015). http://dspace.nbuv.gov.ua/handle/123456789/100645

3. Anderson, J., Peng, F.Z.: Four quasi-Z-source inverters. In: 2008 IEEE Power Electronics Specialists Conference, pp. 2743–2749 (2008). https://doi.org/10.1109/pesc.2008.4592360
4. Vinnikov, D., Roasto, I.: Quasi-Z-source-based isolated DC/DC converters for distributed power generation. IEEE Trans. Industr. Electron. **58**(1), 192–201 (2011). https://pdfs.semanticscholar.org/48e8/7eb70ae697c75b081f06acee18f963a74e06.pdf
5. Khyzhniak, T.A.: Diahnostyka napivprovidnykovykh peretvoryuvachiv iz zastosuvannyam veyvlet-funktsiy m-ichnoho ahrumentu [Diagnostics of semiconductor converters with use of wavelet functions of m-ical group] (2008). http://www.disslib.org/diahnostyka-napivprovidnykovykh-peretvorjuvachiv-iz-zastosuvannjam-vejvlet-funktsiy-m.html
6. Yamnenko, Y.S., Tereshchenko, T.O.: Spectral methods for processing biotelemetrical data. Electron. Commun. **21**(4), 38–43 (2016). https://doi.org/10.20535/2312-1807.2016.21.4.81904
7. Trakhtman, V.A., Trakhtman, A.M.: Osnovy teorii diskretnyh signalov na konechnyh intervalah [Fundamentals of the theory of discrete signals on finite intervals]. Soviet radio, Moscow (1975). https://books.google.com.ua/books?id=vBT_AgAAQBAJ&pg=PA204&lpg=PA204&dq=Трахтман+A+H&source=bl&ots=XqKVjKcXNe&sig=i3n-ghZdm3iVvgZ_lHL7jRlmr5E&hl=en&sa=X&ved=0ahUKEwjvtby4hojbAhXDIpoKHdSiBBgQ6AEIQzAD#v=onepage&q=ТрахтманAH&f=false
8. Sklar, B.: Cifrovaya svyaz. Teoreticheskie osnovy i prakticheskoe primenenie [Digital communication. Theoretical bases and practical application]. Publishing house "Williams," Moscow (2003). http://www.studmed.ru/sklyar-b-cifrovaya-svyaz-teoreticheskie-osnovy-i-prakticheskoe-primenenie_5fb0497bb4c.html
9. Harmuth, H.: Teoriya sekventnogo analiza: osnovy i primeneniya [Theory of Sequential Analysis: Foundations and Applications]. Mir, Moscow (1980). http://www.studmed.ru/harmut-h-teoriya-sekventnogo-analiza-osnovy-i-primeneniya_9fb8cb5a078.html
10. Alyuminievye ehlektroliticheskie kondensatory [Aluminum electrolytic capacitors]. http://www.platan.ru/docs/library/ALCAP_EPCOS.pdf
11. Golembiovsky, Yu.M., Penkov, B.S.: Avarijnye rezhimy preobrazovatel'noj seti, postroennoj na baze invertorov napryazheniya [Emergency modes of the converter network, built on the basis of voltage inverters]. Tech. Electrodyn **3**, 31–34 (2004). Thematic issue "Power Electronics and Energy Efficiency"
12. Klyuev, V.V., Parkhomenko, P.P., Abramchuk, V.E., et al.: Technical diagnostic tools: a reference book [Tekhnicheskie sredstva diagnostirovaniya: spravochnik]. Mechanical Engineering (1989)
13. Bondarenko, V.M., Redkovets, S.N.: Methods for diagnosing electrical circuits [Metody diagnostiki ehlektricheskih cepej], K. (1985)
14. Khyzhniak, T., Tereshchenko, T., Ovsienko, M., Laikova, L.: Diagnostika DC-DC peretvoryuvachiv z kvazi-impedansnoyu lankoyu [Diagnostics of DC-DC converters with a quasi-impedance link], Kiev, vol. 23, no. 2(103), pp. 42–48 (2018). [Microsystems, electronics and acoustics: pratsi III mizhnar. science.-tech. conf. "Smart technologies in energy and electronics – 2018, Lazurne, 21–25 August 2018]
15. Ivakhnenko, A.G.: Induktivnye metody samoorganizacii modelej slozhnyh sistem [Inductive methods of self-organization of models of complex systems]. K.: Science. Opinion, p. 287 (1975)
16. Linnik, Yu.V.: Metod naimen'shih kvadratov i osnovy matematiko-statisticheskoj teorii obrabotki nablyudenij [The method of least squares and the basics of mathematical-statistical theory of processing observations], 2nd edn. (1962)

Generalized Method of Commutation Processes Calculation in High-Frequency Switched-Mode Power Converters

Alexey Gorodny$^{(\boxtimes)}$ ⓘ, Andrii Dymerets ⓘ, Yevhenii Kut ⓘ,
Yurii Denisov ⓘ, and Denisova Natalia ⓘ

Chernihiv National University of Technology, Chernihiv, Ukraine
aleksey.gorodny@gmail.com, andrey.dymerets@gmail.com,
fharse@gmail.com, den7lltd@gmail.com

Abstract. This article presents a generalized method for calculating of switching processes in high-frequency switched-mode power supply converters.

The purpose of the article is to investigate a novel detailed calculation algorithm, which takes into account influence of converter elements parameters on commutation processes. The calculation method of electromagnetic processes is given on example of a boost switched-mode converter driven by pulse-width modulation. An operational cycle of switched-mode converter was divided into intervals, each of which had a linearized model with a specific main transistor switch parasitic parameters. According to a compiled equivalent substitution circuits, the currents and voltages expressions were calculated for corresponding elements of the circuit, and then the power and energy losses in transistor switch were also calculated.

Keywords: Switched-mode · Power converter · Transistor switch ·
Switching interval · Power losses · Quasi-resonant

1 Introduction

In stationary and on-board energy conversion systems, a special attention is paid to energy efficiency of their components. Such systems contain a large number of switched-mode power supplies (SMPS) of different topologies (buck, boost and mixed). In general, SMPS are based on pulse-width modulation (PWM) or pulse-frequency modulation (PFM) [1–4]. The operation frequency of those converters does not exceed of 300 kHz to avoid of dramatically increasing of switching losses. Quasi-resonant pulse converters (QRPC) driven by PFM allow increasing of SMPS operation frequency (up to 10 MHz) and improving their weight-and-size parameters [5–8]. A resonant circuit is included into their power part. Depend on transistor switching condition, QRPC are divided onto zero-current switching QRPC (ZCS-QRPC), and zero-voltage switching QRPC (ZVS-QRPC) [9–12]. Operation of these converters are described in [13–16].

Optimal efficiency parameters of converter can be found by dint of electromagnetic processes analysis taking into account of their nonlinearity that is a difficult task.

© Springer Nature Switzerland AG 2020
A. Palagin et al. (Eds.): MODS 2019, AISC 1019, pp. 71–80, 2020.
https://doi.org/10.1007/978-3-030-25741-5_8

This problem can be solved by simulation of these processes in modern software environments (e.g. MATLAB Simulink). However, as is shown in [17, 18], simulation results often have a large divergence in comparison with experimental results. This is due to imperfection of transistor switch models in simulation environments [19, 20]. In addition, MATLAB Simulink makes it easy to get only a time characteristics of electromagnetic processes in SMPS. Getting the dependences of a certain characteristic on elements parameters of the converter is a high computational complexity task (and takes a lot of simulation time).

Proposed analytical solution of this problem based on generalized method, which is based on replacement of non-linear SMPS model with a linear model at specific intervals of operation cycle, which taking into account parameters of passive power elements, resonant circuit parameters and transistor switch parasitic parameters. In this paper, a novel generalized method for calculation of electromagnetic and energy processes is presented and described in detail. Pro-posed method allows calculating parameters and characteristics of a complex non-linear circuits taking into account passive elements parameters and basic parasitic parameters of switches. Calculations results obtained by this method for ZCS-QRPC with PFM are given in [21–23].

2   Description of the Generalized Method of Commutation Processes Calculation

In different types of SMPS main amount of energy losses (up to 60%) belongs to transistor switches. Modern diodes based on silicon carbide (SiC) and gallium nitride (GaN) have much less parasitic capacitances, less open-state resistance than modern transistors. Therefore, their parasitic parameters may not be taken into account in considered below models. Generally, the precise transistors models are very complex, contain a large number of reactive components, whose parameters are often lacking in documentation. In SMPS which operating at frequencies above of 500 kHz, only MOSFETs and GaNs can be used [21, 22]. In addition, as shown in [21–23], in open-state the MOSFET channel resistance make main influence on commutation processes, but at opening and closing it is the differential channel resistance and drain-source capacitance. Differential resistance is determined from output characteristics of tran-sistor; drain-source capacitance is covered in documentation. In closed state, leakage current is not more than 10 nA, so resistance of transistor can be considered infinitely large.

Taken into account the above mentioned simplifications, SMPS operation cycle can be divided into several switching intervals. The amount of intervals is depending on SMPS type. At each interval, a linearized SMPS model is composed taking into account of passive elements parameters, supply voltage, and transistor switch parasitic parameters. Voltage on capacitances and currents in inductances at beginning of each interval are equal to corresponding values at end of previous interval (according to the commutation laws).

Further calculation of currents and voltages of converter elements occurs by nodal stress method [21–23]. Matrix of conductivities, vector-column of unknown voltages and vector-column of assigning currents are compiled by an equivalent circuit.

$$[Y] \cdot [U] = [J], \tag{1}$$

where $[Y]$ – matrix of conductivities; $[U]$ – vector-column of unknown voltages; $[J]$ – vector-column of assigning currents.

By solving of Eq. (1), the expression is obtained:

$$[U] = [Y]^{-1} \cdot [J], \tag{2}$$

where $[Y]^{-1}$ –matrix inversed to $[Y]$.

By solving of matrix equation, the operator images for all unknown node voltages, which allow finding operator images voltages applied to converter elements and operator image currents through them can be obtained. By using of Inverse Laplace transform, the expressions for voltages applied to converter elements and currents through them in time domain can be also found.

Power losses at any converter element during a single switching interval can be found by integral equation:

$$P_k = \frac{1}{t_k - t_{k-1}} \cdot \int_{t_{k-1}}^{t_k} u(t) \cdot i(t)dt, \tag{3}$$

where k – the number of switching interval; P_k – power losses in k-th interval; t_{k-1} – start of interval; t_k – end of interval; $u(t)$ – expression of voltage change applied to a converter element; $i(t)$ – expression of current change flown through a converter element.

Energy losses at intervals:

$$Q_k = P_k \cdot (t_k - t_{k-1}). \tag{4}$$

Total energy losses in one period of converter operation:

$$Q_{TOTAL} = \sum_{k=1}^{n} Q_k, \tag{5}$$

where n – switching intervals number per one converter period.

Averaged power losses:

$$P_{AVG} = \frac{Q_{TOTAL}}{T}, \tag{6}$$

where T – duration of one period of converter operation.

3   Calculation Example by a Generalized Method

Figure 1 presents a conventional boost-type SMPS converter circuit, where: E – input voltage; L – booster inductance; u_L – inductance voltage; i_L – current flowing through inductance L; VT – transistor switch; u_C – control voltage; u_{VT} – voltage applied to the switch; i_{VT} – current flowing through the switch; VD – reverse diode; C – filtering capacitance; R – resistance load; u_R – output voltage; i_R – output current.

Figure 2 shows time characteristics of parallel SMPS operation. The timing diagram shows basic intervals:

1. t_0-t_1 – energy storage in inductor;
2. t_1-t_2 – transfer energy into a load.

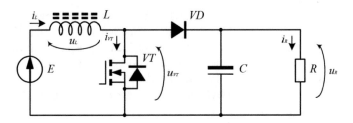

Fig. 1. Scheme of boost SMPS with PWM

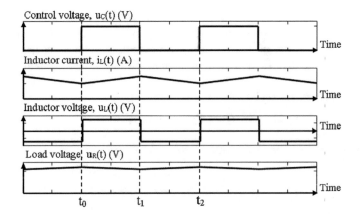

Fig. 2. Time charts of the operation of the boost SMPS

Figure 3 shows switching intervals on transistor switch VT:

1. t_0-t_1 – switch-opening interval;
2. t_1-t_2 – switch open state interval;
3. t_2-t_3 – switch-closing interval;
4. t_3-t_4 – switch closed state interval.

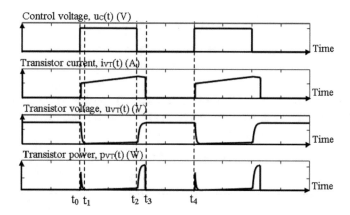

Fig. 3. Time charts of the transistor switch *VT* in boost SMPS

3.1 Transistor Switch Opening Interval (T₀ – T₁)

For transistor switch opening interval the next state-space matrixes are composed (Fig. 4):

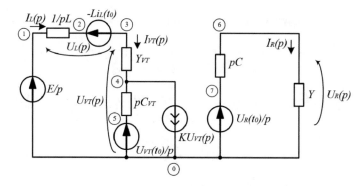

Fig. 4. Operator circuit for switch opening interval

$$[Y] = \begin{bmatrix} Y_{VT} + \frac{1}{pL} & -Y_{VT} & 0 \\ -Y_{VT} + K & pC_{VT} + Y_{VT} & 0 \\ 0 & 0 & pC + Y \end{bmatrix}; \tag{7}$$

$$[U] = \begin{bmatrix} U3(p) \\ U4(p) \\ U6(p) \end{bmatrix}; \tag{8}$$

$$[J] = \begin{bmatrix} \frac{E}{p^2 L} + \frac{i_L(t_0)}{p} \\ u_{VT}(t_0)C_{VT} \\ u_R(t_0)C \end{bmatrix}. \tag{9}$$

As a calculation result of (2) a vector-column $[U]$ consists of three operator node voltage images is obtained. An operator image for all currents and voltages of converter:

$$U_L(p) = \frac{E}{p} - U3(p); \tag{10}$$

$$U_{VT}(p) = U3(p); \tag{11}$$

$$U_R(p) = U6(p); \tag{12}$$

$$I_L(p) = \frac{\frac{E}{p} - U3(p) + Li_L(t_0)}{pL}; \tag{13}$$

$$I_{VT}(p) = (U3(p) - U4(p))Y_{VT}; \tag{14}$$

$$I_R(p) = U6(p)Y. \tag{15}$$

For simplifying of the analytical expressions, the Maple environment were applied.

$$U_L(p) = \frac{pLC_{VT}(Y_{VT}(E - u_{VT}(t_0)) - i_L(t_0)) + L(EKY_{VT} - C_{VT}i_L(t_0))}{p^2 LY_{VT}C_{VT} + p(KLY_{VT} + C_{VT}) + Y_{VT}}; \tag{16}$$

$$U_{VT}(p) = \frac{p^2 LC_{VT}(Y_{VT}u_{VT}(t_0) + i_L(t_0)) + p(LY_{VT}i_L(t_0) + EC_{VT}) + EY_{VT}}{p(p^2 LY_{VT}C_{VT} + p(KLY_{VT} + C_{VT}) + Y_{VT})}; \tag{17}$$

$$U_R(p) = \frac{u_R(t_0)C}{pC + Y}; \tag{18}$$

$$I_L(p) = \frac{Y_{VT}(p^2 LC_{VT}i_L(t_0) + p(KLi_L(t_0) + C_{VT}(E - u_{VT}(t_0))) + EK)}{p(p^2 LY_{VT}C_{VT} + p(KLY_{VT} + C_{VT}) + Y_{VT})}; \tag{19}$$

$$I_{VT}(p) = \frac{Y_{VT}(p^2 LC_{VT}i_L(t_0) + p(KLi_L(t_0) + C_{VT}(E - u_{VT}(t_0))) + EK)}{p(p^2 LY_{VT}C_{VT} + p(KLY_{VT} + C_{VT}) + Y_{VT})}; \tag{20}$$

$$I_R(p) = \frac{u_R(t_0)CY}{pC + Y}. \tag{21}$$

Applying to Eqs. (16–20) the Inverse Laplace transform, the expressions in time domain can be obtained. They in turn allow synthesis of converter timing characteristics, determine effective and maximum values of voltages and currents through the converter elements, calculate power losses in transistor switch.

End of opening interval t_1 can be found from a condition:

$$\left.\frac{du_{VT}(t)}{dt}\right|_{t=t_1} = 0. \tag{22}$$

3.2 Transistor Switch Open State Interval ($T_1 - T_2$)

Electromagnetic processes at open-state interval of transistor switch are calculated in a similar way (Fig. 5).

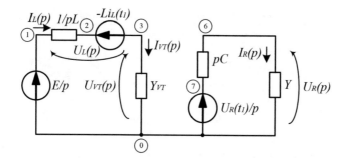

Fig. 5. Operator circuit for switch open-state interval

$$U_L(p) = \frac{L(EY_{VT} - i_L(t_1))}{pLY_{VT} + 1}; \tag{23}$$

$$U_{VT}(p) = \frac{pLi_L(t_1) + E}{p(pLY_{VT} + 1)}; \tag{24}$$

$$U_R(p) = \frac{u_R(t_1)C}{pC + Y}; \tag{25}$$

$$I_L(p) = \frac{pLY_{VT}i_L(t_1) + EY_{VT}}{p(pLY_{VT} + 1)}; \tag{26}$$

$$I_{VT}(p) = \frac{pLY_{VT}i_L(t_1) + EY_{VT}}{p(pLY_{VT} + 1)}; \tag{27}$$

$$I_R(p) = \frac{u_R(t_1)CY}{pC + Y}. \tag{28}$$

3.3   Transistor Switch Closing Interval ($T_2 - T_3$)

$$U_L(p) = \frac{pLC_{VT}(Y_{VT}E - i_L(t_2)) + LY_{VT}(EK - i_L(t_2))}{p^2LY_{VT}C_{VT} + p(KLY_{VT} + C_{VT}) + Y_{VT}};$$ (29)

$$U_{VT}(p) = \frac{p^2LC_{VT}i_L(t_2) + p(LY_{VT}i_L(t_2) + EC_{VT}) + EY_{VT}}{p(p^2LY_{VT}C_{VT} + p(KLY_{VT} + C_{VT}) + Y_{VT})};$$ (30)

$$U_R(p) = \frac{u_R(t_2)C}{pC + Y};$$ (31)

$$I_L(p) = \frac{Y_{VT}(p^2LC_{VT}i_L(t_2) + p(KLi_L(t_2) + C_{VT}E) + EK)}{p(p^2LY_{VT}C_{VT} + p(KLY_{VT} + C_{VT}) + Y_{VT})};$$ (32)

$$I_{VT}(p) = \frac{Y_{VT}(p^2LC_{VT}i_L(t_2) + p(KLi_L(t_2) + C_{VT}E) + EK)}{p(p^2LY_{VT}C_{VT} + p(KLY_{VT} + C_{VT}) + Y_{VT})};$$ (33)

$$I_R(p) = \frac{u_R(t_2)CY}{pC + Y}.$$ (34)

End of opening interval t_3 is can be found from a condition:

$$\left.\frac{du_{VT}(t)}{dt}\right|_{t=t_3} = 0.$$ (35)

Here is an example of calculation of electromagnetic processes obtained by a generalized method for a conventional boost-type SMPS converter driven by PWM (Fig. 6).

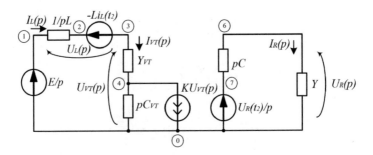

Fig. 6. Operator circuit for switch closing interval

4   Conclusions

- A generalized method allowing calculation of electromagnetic and power circulation processes in SMPS converters with different topologies is proposed;
- A calculation algorithm for currents, voltages and power losses values in all components of converter is showed;
- Proposed method allow using of modelling elements with a different calculation complexity, if their internal parameters are known;
- Proposed method allow evaluating of influence elements parasitic parameters onto energy efficiency of the SMPS converter;
- Qualitative and quantitative assessment results of calculation by proposed method with simulation and experimental results will be given in following articles.

Acknowledgement. This paper was done as a part of the research work "Systems of electric drives with improved energy and dynamic characteristics for special applications" (registration number 0119U000421).

References

1. Shydlovskyi, A.K., Zharkin, A.F., Pazieiev, A.G.: Continuous approximate model of AC/DC converters with active power factor correction. Tekhnichna elektrodynamika **6**, 11–17 (2011)
2. Zhuikov, V.Y., Tereshchenko, T.O., Yamnenko, Y.S., Moroz, A.V.: Adjustable filters of power supplies for information security in microcontrollers: monograph, Kiev, Kafedra, p. 184 (2016)
3. Voytenko, V.P.: Adaptive quasi-optimal control in pulse converters with artificial neural network model of power part. Tech. Electrodynamics **5**, 26–28 (2016)
4. Stepenko, S.A.: Energy efficiency analysis in power factor corrector under different pulse-width modulation modes. Tech. Electrodynamics **4**, 78–80 (2014)
5. Liu, K.: High frequency quasi-resonant converter techniques, Ph. D. Dissertation, Electrical Engineering Department, Virginia Polytechnic Institute and State University, October 1986
6. Zheng, T., Chen, D., Lee, F.C.: Variations of quasi-resonant dc-dc converter topologies. In: IEEE Power Electronics Specialists' Conference Record, pp. 381–392 (1986)
7. Liu, K., Oruganti, R., Lee, F.C.: Resonant switches—topologies and characteristics. IEEE Trans. Power Electron. **2**(1), 106–116 (1987)
8. Tomioka, S., Abe, S., Shoyama, M., Ninomiya, T., Firmansyah, E.: Zero–current–switched quasi–resonant boost converter in power factor correction application. In: 24th Annual IEEE Applied Power Electronics Conference and Exposition (APEC), pp. 1165–1169 (2009)
9. Revko, A.S.: Problem of pulse-width control in quasi-resonant converters. Tekhnichna elektrodynamika **8**, 50–53 (2006). Thematic issue. Problems of modern electrical engineering, Kiev
10. Revko, A.S., Yershov, R.D., Yakosenko, D.S., Beznosko, D.A.: Smooth pulse-width modulation in quasi-resonant pulsed converters using transistor as a voltage-controlled capacitance. In: 2018 IEEE 38th International Conference on Electronics and Nanotechnology (ELNANO), pp. 680–685 (2018). https://doi.org/10.1109/elnano.2018.8477515

11. Revko, A.S., Yershov, R.D., Yakosenko, D.S., Beznosko, D.A.: Stepwise pulse-width modulation in quasi-resonant pulsed converters using switched capacitors. In: 2018 IEEE 38th International Conference on Electronics and Nanotechnology (ELNANO), pp. 711–716 (2018). https://doi.org/10.1109/elnano.2018.8477532

12. Yershov, R.D.: FPGA-based pulse-frequency modulator with adaptive zero-crossing detection for quasi-resonant pulsed converters. In: Proceedings of the 2018 IEEE 38th International Scientific Conference on Electronics and Nanotechnology (ELNANO 2018), Kyiv, Ukraine, pp. 721–726, 24–26 April 2018

13. Voytenko, V., Stepenko, S., Velihorskyi, O., Chakirov, R., Roberts, D., Vagapov, Yu.: Digital control of a zero-current switching quasi-resonant boost converter. In: Proceedings of Sixth International Conference on Internet Technologies & Applications, Wrexham, North Wales, UK, pp. 365–369, 8–11 September 2015

14. Denisov, Y.O., Stepenko, S.A., Gorodny, A.N., Kravchenko, V.A.: Input current parameters analysis for pfc based on quasi-resonant and conventional boost. In: IEEE Thirty-Fourth Annual International Scientific Conference on Electronics and Nanotechnology (ELNANO), pp. 393–397 (2014)

15. Gorodniy, O., Gordienko, V., Stepenko, S., Boyko, S., Sereda, O.: Impact of supply voltage change on the energy performance of boost quasi-resonant converter for radioelectronic equipment power supplies. In: Modern Electrical and Energy Systems (MEES), pp. 232–235 (2017)

16. Denisov, Yu., Gorodny, A., Gordienko, V., Vershniak, L., Dymerets, A.: Estimation of parameters and characteristics of the factor correction factor based on pulsed and quasi-resonant power converters. Tekhnichna elektrodynamika **6**, 38–41 (2018)

17. Voytenko, V., Stepenko, S.: Simulation peculiarities of high-frequency zero-current switching quasi-resonant boost converter. In: Proceedings of 2015 IEEE 35th International Conference on Electronics and Nanotechnology (ELNANO), pp. 486–491, 21–24 April 2015

18. Ivakhno, V., Zamaruiev, V., Ilina, O.: Estimation of semiconductor switching losses under hard switching using Matlab/Simulink subsystem. Electr. Control Commun. Eng. **1**(2), 20–26 (2013)

19. Li, G.: An improved MATLAB/simulink model of SiC power MOSFETs. In: 2016 IEEE 8th International Power Electronics and Motion Control Conference (IPEMC-ECCE Asia). IEEE (2016)

20. Sun, K.: Modeling of SiC MOSFET with temperature dependent parameters. In: Zhongguo Dianji Gongcheng Xuebao (Proceedings of the Chinese Society of Electrical Engineering), vol. 33, no. 3 (2013)

21. Denisov, Y., Gorodny, A., Gordienko, V., Yershov, R., Stepenko, S., Kostyrieva, O., Prokhorova, A.: Switch operation power losses of quasi-resonant pulse converter with parallel resonant circuit. In: IEEE Thirty-Fourth Annual International Scientific Conference on Electronics and Nanotechnology (ELNANO), pp. 327–332 (2016)

22. Denisov, Y., Gordienko, V., Gorodny, A., Stepenko, S., Yershov, R., Prokhorova, A., Kostyrieva, O.: Power losses in MOSFET switch of quasi-resonant pulse converter with series resonant circuit. In: 2nd International Conference on Intelligent Energy and Power Systems (IEPS), pp. 1–6 (2016)

23. Denisov, Yu., Gorodny, A., Gordienko, V., Stepenko, S., Yershov, R., Tepla, T.: Comparison of power losses in switches of increasing number of common-state directives with parallel and series resonant circuits. Tekhnichna elektrodynamika **4**, 44–46 (2016)

Mathematical Modeling and Simulation of Systems in Information Technology and Information Security

Traffic Abnormalities Identification Based on the Stationary Parameters Estimation and Wavelet Function Detailization

Nikolai Stoianov[1] , Vitalii Lytvynov[2] , Igor Skiter[3]([⊠]) ,
and Svitlana Lytvyn[2]

[1] Bulgarian Defence Institute,
2 Prof. Tsvetan Lazarov Blvd, 1592 Sofia, Bulgaria
n.stoianov@di.mod.bg
[2] Chernihiv National University of Technology,
Shevchenko str. 95, Chernihiv 14035, Ukraine
v.vlytvynov.dept@gmail.com, chdtu.fld@gmail.com
[3] Institute of Mathematical Machines and Systems Problems, National Academy
of Science of Ukraine, Ac. Glushkov Avenue 42, Kyiv 03680, Ukraine
skiteris@ukr.net

Abstract. The article deals with the analysis of the of corporate network traffic properties using statistical methods. The complex method of estimating the stationary of computer network traffic using the autocorrelation function and the Hurst index is proposed. To detect the abnormal behavior of computer network traffic, the wavelet analysis method using the Fourier function, has been developed and tested.

Keywords: CyRADARS · Autocorrelation · Hurst index · Persistence · Wavelet analysis · Detailed fourier function

1 Introduction

The NATO Science for Peace and Security Programme includes the «Cyber Rapid Analysis for Defence Awareness of Real-time Situation – CyRADARS» project [1, 17]. Within the framework of this project, the Information Security Training Center (ISTC) was established.

The objectives of the center are modeling, analysis and decision making on the security of corporate information networks. The structure of the center includes laboratories for attack simulation (ASL), an attack defense lab (ADL) and an analytical research laboratory (ARL).

The main task of the ARL is to develop methods for detecting behavior. In addition, the goals are set: classification of algorithms and their capabilities; research areas of attack detection algorithms and non-standard behavior algorithms effective use; research areas of non-standard network behavior algorithms effective use; development and modification of algorithms for detection of NSNBs, with a high level of efficiency;

© Springer Nature Switzerland AG 2020
A. Palagin et al. (Eds.): MODS 2019, AISC 1019, pp. 83–95, 2020.
https://doi.org/10.1007/978-3-030-25741-5_9

development of priority control algorithms for the use of NSNBs detection algorithms using information on monitoring the security of regional infrastructure.

The exploitation of corporate information systems can be subjected to various external influences: attacks, intrusion, technical malfunction, unauthorized interference, etc. All these phenomena can lead to negative consequences of the network operation and can be called abnormal behavior.

The object of research is the corporate network of the Faculty of Electronic and Information Technologies (FEIT) of the Chernihiv National University of Technologyes (CNUT).

The task of the first stage of the research is to create a method for the qualitative evaluation of network traffic abnormal behavior. The methodology should meet the following characteristics: efficiency, adequacy, productivity, low consumption of information resources.

Detection of anomalies in the corporate network can be realized under the condition of determining the idealized network behavior - "network profile". Then, by comparing real and ideal traffic, you can estimate abnormal properties of the network in real time.

2 Statistical Methods for Development and Modification of Algorithms for Non-standard Behavior Detection, Which Have a High Level of Efficiency

The analysis of computer network traffic using statistical methods includes the requirement to estimate the reference of idealized network behavior. Ideally, one can consider the state of the network in the absence of anomalies in it. The use of statistical methods to detect the non-standard network behavior of traffic is based on a comparison of the ideal and real behavior of traffic. In this case, the problem of comparison is reduced to the deviations levels analysis of data dynamic range, their series, etc. Unlike signature methods, statistical network analysis methods do not require information on the characteristics and parameters of network intrusions at previous intervals.

The advantages of statistical methods include the fact that they can be modified, adapted and used in a complex depending on the characteristics of the object analysis. In addition, statistical methods can detect new methods of impact on the network. The disadvantages of statistical methods are the following: the impossibility of identifying factors that affect the occurrence of anomalies, errors of the first and second kind.

Computer network traffic is a dynamic data set. It can be represented as an additive or multiplicative function (1)

$$\begin{cases} Y(t_i) = tr(t_i) + S(t_i) + A(t_i) + e(t_i) \\ Y(t_i) = tr(t_i) \times S(t_i) \times A(t_i) \times e(t_i) \end{cases} \tag{1}$$

where $tr(t_i)$ - is the general trend; $S(t_i)$ - is the periodic component (cyclic, seasonal), defined for the selected interval; $A(t_i)$ - are the anomalies, local characteristics of the traffic behavior, sharp changes; $e(t_i)$ - is the stochastic component, noise.

The use of statistical methods for analyzing network traffic is based on the determination of traffic components and their behavior. The analysis of statistical methods used in the estimation of network traffic abnormal behavior is represented in articles [2, 3].

Determination of the general trend of the dynamic series is carried out using polynomial models, exponential models [4] and logistic models [5]. The choice of the trend model type depends on the behavior of the dynamic series levels and their change over time.

Simulation of the cyclic (seasonal) component of a dynamic series is usually carried out using harmonic analysis [6].

The parametric uncertainty of dynamic data series describing network traffic is determined by the trend and cyclic component. Therefore, the use of traditional modeling trends and seasonality leads to the inadequacy of these models. The elimination of this disadvantage is possible when using the theory of wavelet analysis.

In the article [7, 8] the authors describe the basic wavelet analysis algorithms for the problem of detecting anomalies in networks and compare them. The advantage in modeling and network traffic is provided by an algorithm based on discrete wavelet analysis of network traffic. The peculiarity of the algorithm use is the ability to analyze the traffic parameters in the selected "study window". The sampling of traffic is carried out on an interval Δt_i. In the process of research, it can be changed $\Delta t_i \pm \Delta t'$, ($\Delta t_i < \Delta t'$) This approach makes it possible to detect anomalies not found in previous researches.

The wavelet model has the following form:

$$Y(t_i) = \sum b_{k,\tau} \varphi_{k,\tau}(t_i) + \sum d_{k,\tau} \omega_{k,\tau}(t_i); \quad k, \tau = \overline{1, \infty} \tag{2}$$

where $\varphi_{k,\tau}(t_i)$ - is the scaling function, network traffic approximation function; $\psi_{k,\tau}(t_i)$ - is the wavelet function, network traffic details, and its local features; $b_{k,\tau}, d_{k,\tau}$ - are the approximating and detailing coefficients with scale k and offset τ parameters.

Its use, subject to the choice of adequate approximating and detailed functions, will allow to estimate wavelet parameters. Defined estimates of the parameters of the traffic windows form their own array of statistical data. Investigation of properties of this massif parameters estimations, as well as the determination of the law of their distribution, will allow to estimate possible abnormalities and determine the time of their occurrence.

The first part of Eq. (2) simulates the general trend (tendency) and cyclicity. The second part describes the fluctuations of traffic in the intervals of the sample. They describe the magnitude of the deviation or abnormality of the subjects network taking into account the stochastic component.

The research made by the authors [6–8] showed that the quality of detecting abnormal behavior of traffic depends on the type of scaling function and wavelet type. For example, it is proposed to use the - Fourier function as well as the Haar wavelet as a scaling function.

Determining the abnormal behavior of traffic in real time is possible using predictive models. In this case, an adequate traffic model can be obtained using the described wavelet analysis method.

An effective method for analyzing the behavior of dynamic series and predicting it for subsequent periods are autoregressive models [9, 10]. They represent the models that have a close correlation between the next and previous values of the series. The main class of autoregressive models used in effective prediction of dynamic series includes:

– Autoregressive model $AR(p)$:

$$Y(t) = const + \sum_{\tau=1}^{p} \alpha_\tau Y_{t-\tau} + e_t \tag{3}$$

where $Y(t)$ - is the dynamic range element; $const$ - is the constant of dynamic range; $\alpha_\tau, \tau = \overline{1,p}$ - is the autoregression parameter estimates; τ - is the lag; e_t - is the stochastic component, noise.

– Autoregressive moving-average model, ARMA (p,q):

$$Y(t) = const + \sum_{\tau=1}^{p} \alpha_\tau Y_{t-\tau} + \sum_{i=1}^{q} \beta_i Y_{t-\tau} + e_t \tag{4}$$

where $\beta_i, i = \overline{1,q}$ - moving average parameters estimates; q - moving average order.

– Autoregressive integrated moving average ARIMA (p, Δ, q):

$$\Delta^d Y(t) = const + \sum_{\tau=1}^{p} \alpha_\tau \Delta^d Y_{t-\tau} + \sum_{i=1}^{q} \beta_i Y_{t-\tau} + e_t \tag{5}$$

where Δ^d- is the dynamic range difference operator size d.

As traffic prediction models, researchers [11] used a group of machine learning methods and a group of statistical methods. The following methods were used in this research: Support Vector Machines (SVM), Radial Basis Function (RBF) Neural Network, Multilayer Perceptron (MLP), M5P (a decision tree with linear regression functions at the nodes), Random Forest (RF), Random Tree (RT), and Reduced Error Pruning Error (REPE), and a statistical regression called Holt-Winters have been used to predict the amount of network traffic in TCP/IP-based networks Internet Service. The effectiveness of prediction models for the data sets was estimated using the Mean Absolute Percentage Error (MAPE).

Also in [6] other methods for predicting and modeling time series were considered for the best use of network traffic data to increase cybersecurity. Researches have shown that before applying statistical methods to raw network data, it is necessary to perform substantial preliminary processing. In addition, logistic regression is the most effective for predicting. Also, the advantage of the model is that it can be used to solve classification problems.

However, the researchers did not evaluate the adequacy of the models. Therefore, the question of choosing the right model is not resolved. Standard criteria were used as comparison criteria, such as the average proportion of correctly defined classes (in classification problems) and the mean absolute error (in row-level prediction problems). It should also be noted that when constructing the models, such traffic properties as persistence, seasonality, and self-similarity are taken into account.

In the researches of scientists [12], it was concluded that the prediction dynamics of network traffic has greater reliability and adequacy in the use of nonlinear models. In addition, it is noted that logistic models and autoregressive models are more efficient. It is concluded that the choice of the model class for prediction depends on the interval of traffic sampling.

The ARIMA model for analyzing network traffic is commonly used to predict network traffic. In [13] the algorithm for applying the ARIMA model includes a dynamic error correction coefficient having algorithm, based on the principle of predicting linear minimum variance. This can solve the problem of stochastic effects in prediction, and the prediction accuracy improves.

Traffic analysis using statistical methods will not be complete unless its properties of similarity are studied. Self-similar models may have a long-term property. This indicates the existence of a correlation between events arrays over long periods of time. When analyzing network traffic, the presence or absence of long-term dependencies in dynamic data series has a close relationship with analytical forecast models and the possibilities of their use to defect traffic anomalies in real-time.

Real series of network traffic dynamics contain a large amount of data. They also have deterministic and stochastic properties. Therefore, we can talk about their fractal nature. To characterize the fractal structure, the fractal Hausdorff index D is used [14]. To characterize the dynamics, the following properties are considered: persistence (resistance to the trend), antipersistence ("return to the average"), cyclicity. The indicator is closely related to the Hurst index [15], which determines the properties of the system dynamics or dynamic series. For a Gaussian process, the Hurst index is defined as:

$$H = 2 - D \tag{6}$$

At traffic discretization, the definition of the Hurst index is performed using R/S-analysis. For most dynamic series, the following equality holds:

$$\frac{R}{S} = (a \times N)^H \tag{7}$$

where R/S - normalized range of accumulated average; S - standard deviation; a - constant for each specific process; N - number of observations; H - Hurst index $(0 < H < 1$, characterizes the fractal dimension of the process.

The algorithm for calculating the index is given in [16] and calculated by (8):

$$H = \frac{\log\left(\frac{R}{S}\right)}{\log a + \log N} \tag{8}$$

Hurst index is used to characterize the dynamic range and calculate the correlation coefficient r (9):

$$r = 2^{2H-1} - 1 \qquad (9)$$

Equation (9) is used to estimate the effect of autocorrelation of previous values of the dynamic range on its next values and determine the future trend. Depending on the values H, r the following dynamic range properties are classified:

1. If $0,5 < H < 1$ ($1 < D < 1,5$) dynamic range trend-resistant; (the higher the value of the indicator r - the higher the indicator H).
2. If $H = 0,5$ ($D = 1,5$) the state of the system is random absolutely independent without any correlation ($r = 0$): the current state of the system is in no way related to its future state.
3. If $H < 0,5$ ($D > 1,5$) - "Antipersistent range"; the forecast trend is opposite to the previous period.

Thus, we can draw the following conclusion. The analysis of computer network traffic using statistical methods includes the requirement to evaluate the reference or idealized network behavior.

Ideally one can consider the state of the network in the absence of anomalies in it. The use of statistical methods to detect the non-standard behavior of traffic in the network is based on a comparison of the ideal and real behavior of traffic. In this case, the problem of comparison is reduced to the analysis of deviations of the levels of data dynamic range, their series, etc. Unlike signature methods, the methods of statistical network analysis do not require information about the characteristics and parameters of the network.

The advantages of statistical methods include the fact that they can be modified, adapted and used in a complex depending on the characteristics of the object analysis. In addition, statistical methods can detect new methods of impact on the network. The disadvantages of statistical methods can be considered: the impossibility of identifying factors that affect the anomalies occurrence, errors of the first and second kind.

3 Research Algorithm

The previous section describes some statistical methods for analyzing the traffic. According to the authors, they can be used in a complex to estimate the state of network traffic and detect anomalies in it. Figure 1 shows a block diagram of a comprehensive statistical analysis of network traffic. It is based on the use of three main methods: wavelet analysis, auto-regressive analysis and elements of fractal analysis.

Analysis of network traffic using wavelet analysis is carried out in [8]. As a function of scaling, the Fourier function is selected. The algorithm for determining the parameters of approximating and detailing functions is shown in Fig. 2.

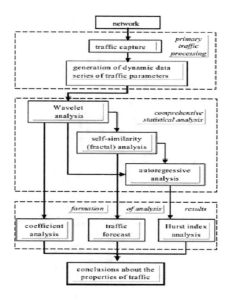

Fig. 1. Block diagram of a comprehensive statistical analysis of network traffic

The use of the Fourier series as an approximation function gives the opportunity to obtain the most accurate results and determine the periodic components of the network traffic. When analyzing traffic, it is important that the Fourier function clearly expresses the amplitudes and the initial phases of the harmonic components.

As a result, you can establish the similarity or difference between different sections of traffic in different conditions, usually when comparing the "ideal profile" with the real traffic. The process of selecting the wavelet that most accurately reflects the actual traffic is iterative and is performed for $k = 1$, $k = 2$, $k = 3$, ... with the definition of the approximation error at each step. Stopping criterion is the minimum value of the approximation error σ_k.

Determination of the parameters of detailing and approximating functions can be carried out under different conditions.

For example: using the method of moving average order q, increasing the size of the wavelet, we sample the traffic at different values of the Δt. As a result, we obtain a statistical array of approximating and detailing coefficients. A significant change in their absolute value or inaccuracy of statistical tests in terms of significance, efficiency, and validity make it possible to conclude that the anomalies of the investigated traffic.

In addition, the wavelet analysis allows to synthesize an adequate model of the behavior of network traffic. Thus we obtain the estimates of the function $\hat{Y}_t$. Deviation of real traffic Y_t from its idealized model can serve as an indicator of abnormal behavior at certain moments of time t.

The next step in analyzing network traffic is to determine its deterministic/ stochastic properties using the Hurst index. In the case when the sample size is a significant amount (several thousand positions), it is advisable to use the modification

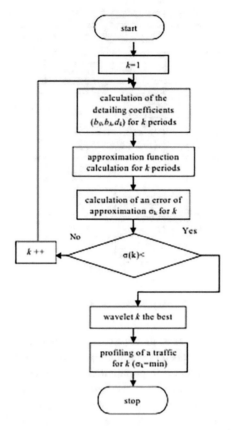

Fig. 2. Algorithm for determining the parameters of the forming function during wavelet analysis

of the proposed algorithm presented in [17]. According to the algorithm, the Hurst index is defined as the slope tangent in linear regression:

$$\log\left(\frac{R}{S}\right)_{n_t} = f(\log(n_t)) \tag{10}$$

where n_t - is the number of discrete time periods

But the classical Hurst method has several drawbacks. One of them is the impossibility of calculating the indicator in real time due to the large volume of calculations. One of the options for eliminating this disadvantage may be the use of a step-by-step recurrent algorithm:

$$H(t+k) = \frac{\log\left(\frac{R(t+k)}{S(t+k)}\right)}{\log(t+k) + \log a} \tag{11}$$

where k = 1, 2, ... - are the time intervals of aggregation of observation data on the process of real network traffic. From the expression (11) it follows that the Hurst index can be refined at each stage of aggregation without first memorizing the values of the data stream. Determining the Hurst parameter for real and discretized network traffic will make it possible to determine its prognostic properties and the level of stationary/non-stationary more accurately.

The autoregressive analysis of network traffic includes the following steps:

1 -   determination of the autoregression model;
2 -   determination of the order and parameters of regression;
3 -   check in the adequacy of the model;
4 -   prediction.

Research at step 1–2 chow that the ARIMA (p, d, q) model, an autocorrelation function (ACF) or a specific ACF is determined. The parameter p is determined by the minimum value of the Mean Square Error (MSE), parameter d - by bringing the series to the stationary (using the successive difference operator), the parameter q- is selected at the stage of conducting the wavelet analysis (the order of the moving average).

Step 3 shows, the verification of the adequacy of the error analysis that is performed using the Ljung–Box test to distinguish whether any of the autocorrelation group of a time series are different from zero:

$$Q = n(n - 2) \sum_{i=1}^{\tau} \frac{r_i^2}{n - \tau} \tag{12}$$

where n - is the sample size; τ - lag; r_i - is the sample autocorrelation.

Step 4, the possibility of using the model for predictiry is estimated using Mean Absolute Percent Error (MAPE).

4   Experiment Technique and the Results of Experimental Researches

The data array for network traffic analysis is based on data network traffic packets (bytes) of the Faculty of Electronic and Information Technologies (FEIT) of Chernihiv National University of Technology (CNUT). Observation period: 08.01.2019–12.01.2019; 8:00 AM – 8:00 PM; data aggregation time- 1 min, number of observations–3600 values. To simulate the wavelet function, data for 1 day (720 values) are used (Fig. 3), as well as selective traffic with $\Delta t = 10$ min (Fig. 4). Wavelet size is 2 h. The implementation of the wavelet model using the Fourier approximation function was carried out with a lag $\tau = 2$ (20 min). As a result of simulation, the coefficients of approximation $b_{k,\tau=2}$ are obtained (Fig. 5).

Results simulation of real network traffic of FEITs with help of wavelet analysis with Fourier approximation function is shown in Fig. 5.

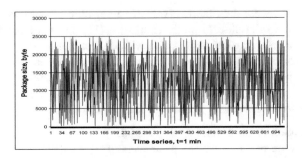

Fig. 3. Network traffic the packet size (byte) the network FEIT of the Chernihiv National University of Technology (CNUT) per 08/01/2019.

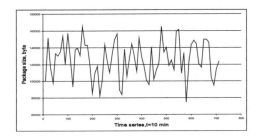

Fig. 4. Network traffic the packet size (byte) the network FEIT of the Chernihiv National University of Technology (CNUT) per 08/01/2019.

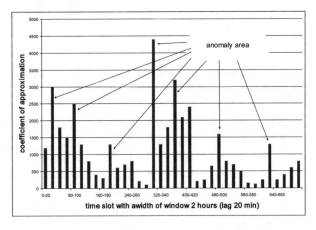

Fig. 5. The coefficients of the approximation of the wavelet function network traffic of the period 08/01/2019.

On the basis of the simulation results (Fig. 6), the parameters for constructing the Eq. (10) are determined. The Hurst index for the indicated dependence is H = 0.5697 (Fig. 7). This means that the network traffic that has been studied has persistent properties, is trend-resistant.

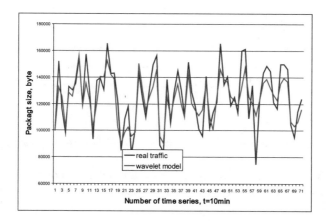

Fig. 6. Wavelet analysis network traffic of FEITs with Fourier approximation function of the period 08/01/2019.

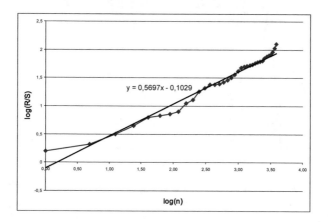

Fig. 7. Calculation of the Hurst index

The ARIMA (p, d, q)-model was chosen to construct an autoregression model. The parameter p = 12 is defined as the last coefficient of a partial autocorrelation function, which is significantly different from zero (Fig. 7). Parameter q = 3 is defined as the last coefficient of an autocorrelation function, (Fig. 8).To determine the parameter Δ^d, the successive difference operator was applied twice, hence $\Delta^d = 2$.

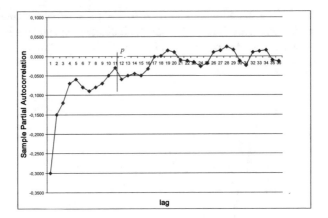

Fig. 8. Calculation of the parameter *p* for ARIMA (p, d, q)-model

Prognostic properties of the model were estimated at approximately numerically equal MAPE = 18.3%. This means that the predictive properties of the model are very good (Fig. 9).

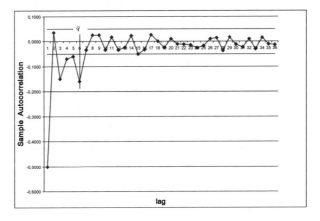

Fig. 9. Calculation of the parameter *q* for ARIMA (p, d, q)-model

5 Conclusions

The method of identifying abnormal network traffic with a set of statistical methods is proposed. On the basis of the modified wavelet function, an estimate of the traffic anomalies was carried out for a sharp change in the estimates of the parameters of the approximation function. The suitability of using the model to estimate traffic anomalies for predicting periods, the stability of its trend was evaluated using the Hurst index. For an adequate prediction, the ARIMA (p, d, q)-model was constructed. Model parameters are determined based on an analysis of correlation coefficients. The quality of the model is very good. It can be used to predict traffic for future periods.

Acknowledgment. This research is funded by the NATO SPS Project CyRADARS (Cyber Rapid Analysis for Defense Awareness of Real-time Situation), Project SPS G5286.

References

1. NATO SPS Project CyRADARS (Cyber Rapid Analysis for Defense Awareness of Real-time Situation). https://www.cyradars.net
2. Haviluddin, H., Alfred, R.: Forecasting network activities using ARIMA method. J. Adv. Comput. Netw. (JACN) **2**(3), 173–179 (2014)
3. Purnawansyah, P., Haviluddin, H., Alfred, R., Gaffar, A.F.O.: Network traffic time series performance analysis using statistical methods. Knowl. Eng. Data Sci. (KEDS) **1**(1), 1–7 (2018)
4. Bojilov, T.: Classification and Analysis of Computer Network traffic. Ph.D. Thesis. Networking & Security Department of Electronic Systems Aalborg University, p. 262 (2014). www.es.aau.dk/netsec. ISBN:978-87-7150-30-9
5. Major Jr., W.F.: Statistical Analysis of the Skaion Network Security Dataset. Ph.D. Thesis, Naval Postgraduate School Monterey, p. 55 (2012). CA 93943-5000
6. Bernacki, J., Kołaczek, G.: Anomaly detection in network traffic using selected methods of time series analysis. Int. J. Comput. Netw. Inf. Secur. **9**, 10–18 (2015)
7. Shelugin, O.I., Pankrushin, A.V.: Measuring the reality of detecting network anomalies using discrete wavelet analysis. In: Science and Information (SAI), Conference, London, UK, pp. 393–397 (2015)
8. Komorowski, D., Pietraszek, S.: The use of continuous wavelet transform based on the fast fourier transform in the analysis of multi-channel electrogastrography recordings. J. Med. Syst. **40**, 10 (2016). https://doi.org/10.1007/s10916-015-0358-4
9. Box, G.E.P., Jenkins, G.M.: Time Series Analysis: Forecasting and Control. Holden-Day, San Francisco, p. 575 (1970–1976)
10. Brockwell, P.J., Davis, R.A.: Introduction to Time Series and Forecasting, 2nd edn, p. 449. Springer, New York (2002)
11. Akgol, D., Akay, M.F.: Network traffic forecasting using machine learning and statistical regression methods combined with different time lagss. Int. J. Adv. Electron. Comput. Sci. **3**(10), 68–72 (2016). ISSN: 2393-2835
12. Lukas, K.M., Solanic, Y.V., Ovchinnikiv, K.A., David, O.O.: Comparative analysis of traffic forecasting methods in telecommunication networks. Electron. Sci. Publ. J. Probl. Telecommun. **1**(13), 84–95 (2017). http://pt.jornal.kh.ua)
13. Wang, J.: A process level network traffic prediction algorithm based on ARIMA model in smart substation. In: IEEE International Conference on Signal Processing, Communication and Computing (ICSPCC 2013), KunMing, pp. 1–5 (2013)
14. Hausdorff, F.: Dimesion und ausseres mass. Matematishe Ann. **79**, 157–179 (1919)
15. Hurst, H.E.: A long-term storage capacity of reservoirs. Trans. Am. Soc. Civ. Eng. **116**, 770 (1951)
16. Skiter, I.S., Trunova, O.V.: Algorithmization process for fractal analysis in the chaotic dynamics of complex systems structured type. Int. J. Inf. Theor. Appl. **22**(3), 277–285 (2015)
17. The NATO Science for Peace and Security (SPS) Programme. Annual Report 2017. 60th Anniversary Special Edition, p. 92

Development of Model for Load Impacts to the Database

Oleksandr Khoshaba[1]([⊠]) [iD], Vitalii Lytvynov[2][iD], and Artem Zadorozhnii[2][iD]

[1] Institute of Mathematical Machines and Systems Problems of the Ukraine National Academy of Science, 42 Academician Glushkov Avenue, Kyiv 03187, Ukraine
Oleksandr.Khoshaba@gmail.com
[2] Chernihiv National University of Technology,
95 Shevchenka Str., Chernihiv 14035, Ukraine
v.v.lytvynov.dept@gmail.com, zaotroy@gmail.com

Abstract. The article describes some of the results of database performance studies, identifies features of the influence of some factors on their functioning. The wording of the problem is fulfilled and the ways of their solution are indicated. Also, a model of the element as the main component of the load impacts is proposed. The input and output parameters of the element for load impacts are shown, and their characteristics are provided. An example of load impacts using the software tool for testing open MySQL database connections is given.

Keywords: Models of load impacts · Perfomance testing · Database performance

1 Introduction

It's very important that start load testing early in the software development life cycle, on the developer's machine. QA and performance engineering teams can run large load tests in the server later in the development cycle. It's need catch website, app and API performance problems before production. Testers use the same test scripts across local and server execution modes. Also, traditionally, interpreting performance test results has been difficult in any time. The study of load impacts of query flow results makes it easy to understand all system test results. So, users can quickly find and fix computer performance issues.

Performance testing is a form of software testing that focuses on how a system works under a particular load. Performance testing does not search for software errors or defects. Performance testing should use reference scripts, standards, and software tools. As a result of performance testing, diagnostic information is provided necessary to identify and eliminate bottlenecks in the computer system.

2 Analysis of Research and Publications

In practice, various studies are carried out [1–8] the impacts of database query flows. So, there are many factors that have a significant impacts on database

© Springer Nature Switzerland AG 2020
A. Palagin et al. (Eds.): MODS 2019, AISC 1019, pp. 96–102, 2020.
https://doi.org/10.1007/978-3-030-25741-5_10

queries. Such factors include [1] the presence of developed equipment, a high level of parallelism and blocking of the storage mechanism (blocking of tables and rows). The work also notes that this indicator is useful for finding queries that most affect the response time of an application or load the server most [1].

In other papers [2–4] it is noted that the operation of deleting a large number of rows from the database represents a rather large load for which special methods need to be used: reduce the size of transactions, increase the time between DELETE operators to reduce the strength of requests.

In [5], the SQL Server 2016 was tested where the AutoAdjustBufferSize parameter was investigated. The AutoAdjustBufferSize parameter influenced the strength of the database query flow. Changing the AutoAdjustBufferSize parameter improved database performance [5].

In [6] and [7], models were obtained and the results of performance studies of object-relational mapping (ORM) frameworks with databases were obtained. As a result of these studies [6] and [7], various types of structural queries were considered.

In [8], the importance of developing models that affect the performance of databases is noted, but methods and quantitative assessments of their creation are not described.

3   Problem Formulation

Thus, although the problem of studying the load impacts of the flow of requests has existed for a long time, there are still questions related to their measurement. In this article, we propose a model for load impacts to the database that solves the following problem.

There is a database located on the computer network server. In order to conduct performance testing to the database from client workstations receive requests. The rate of requests is controlled by written script. Also, the rate of requests is a function. It is necessary to create a load impacts model for the database.

4   The Solution of the Problem

4.1   The Model of Element for Load Impacts

Regulation of the flow of requests to the database is an important aspect of the study of load impacts. A script is used to regulate the flow of requests. Therefore, the basis of the load impacts is the scenario. The scenario consists of individual elements of the load impacts, which have input and output parameters (Fig. 1).

Input parameters of the load element (Fig. 1) are: rate of requests (RR) and total number of requests (NQ). Output parameters of the load element are: Total Time (TT) and Number of Stage (NS).

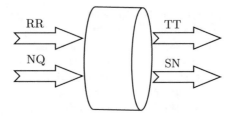

Fig. 1. The model of element for load impacts

4.2   Quantitative Characteristics of the Rate of Requests

The rate of requests is an input parameter of element for load impacts (Fig. 1). Based on the request flow, the following quantitative characteristics of the rate of requests are: rate of change requests (RCR) and total query rate (TQR).

If rate of requests (RR) denotes as $f'(t)$, then rate of change requests (RCR) denotes as $f''(t)$.

Than, these characteristics have following rations for the rate of requests (RR):

$$RR = \int f''(t)dt; \tag{1}$$

$$RR = \frac{dTQR}{dt}; \tag{2}$$

For the rate of change requests (RCR):

$$RCR = \frac{dRR}{dt}; \tag{3}$$

Also, we can define total query rate (TQR) as:

$$TQR = \int f'(t)dt; \tag{4}$$

In some cases, we find the derivative of RCR - $DRCR$. Then:

$$DRCR = \frac{dRCR}{dt}; \tag{5}$$

Also:

$$RCR = \int f^{(3)}(t)dt; \tag{6}$$

There are also evaluations of the characteristics of load impacts which are as follows for total query rate (TQR):

$$evTRQ = \int_{t_0}^{t_1} RR(t)dt \tag{7}$$

For rate of requests (RR) we can define:

$$evRR = \int_{t_0}^{t_1} RCR(t)dt \tag{8}$$

For rate of change requests (RCR) we can define:

$$evRCR = \int_{t_0}^{t_1} DRCR(t)dt \tag{9}$$

4.3 Qualitative Characteristics of the Load Impacts

The stage is the output parameter of element for load impacts (Fig. 1). The stage has gradations, which, depending on the input parameters of element for load impacts. During load impacts, the object of study can take on various stages.

Stages have gradations and are measured in units. We can designate them as conditions S0, S1 and S2. These conditions can be referred to as low, moderate (medium) and high loads. The criteria for assigning a state to gradations can be the presence of a queue, the appearance of errors, the loss of requests, the synchronous or asynchronous transmission of requests.

For example, criteria such as the execution of synchronous single requests, the absence of a queue, errors, and lost requests are related to the S0 stage.

The stage S1 of the object of study includes such criteria as the multiple asynchronous requests, presence of a queue, but the absence of errors and lost requests.

The stage of S2 of the object of study includes such criteria as the presence of a queue, but the absence of errors and lost requests, multiple asynchronous requests.

In some cases, transitional gradations between the stages S1 and S2 can be determined.

4.4 An Example of Determining the Characteristics of the Load Impacts

The scenario has a function that is used by the elements for the load impacts. For example, we define the flow to database queries $RR(t)$ by function (Fig. 2):

$$RR(t) = 3,7t - 0,3t^2[*10^2 rps]; \tag{10}$$

The time period will be from 0 to 12 s. Next, we define the functions RCR (3) as:

$$RCR(t) = RR'(t) = 3,7 - 0,3t; \tag{11}$$

Then, we define the functions TRQ (4) as:

$$TQR(t) = \int RR(t)dt = -\frac{t^2(2t - 37)}{20} + C; \tag{12}$$

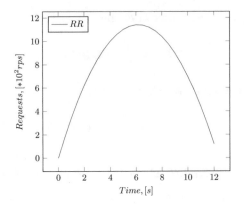

Fig. 2. The quantitative characteristics of the load impacts

The functions of the quantitative characteristics of the load impacts for $RCR(t)$ (11) and $TQR(t)$ (12) functions are shown in the Figs. 3 and 4.

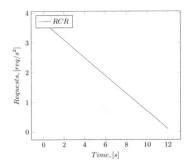

Fig. 3. Rate of change requests function

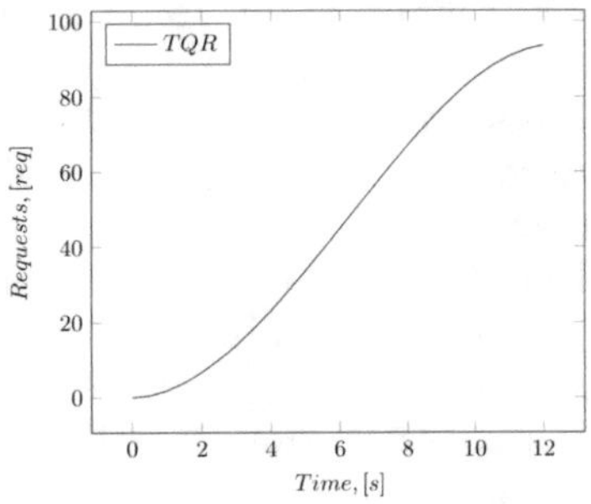

Fig. 4. Total query rate function for $C = 0$

This study was performed to examine the performance of open MySQL database connections using a server Dell Poweredge R620 (2 x Intel Xeon E5-2690 Processors).

The load impact scenario was written in a bash script which consisted of the following command:

nping -c 1000 --rate n --tcp -p 3306 coll

where

c - number of requests;
rate n, for example 1100 - rate of requests;
--tcp -p 3306 - open connection port for MySQL database.

This command is used as a software tool for performance testing of open MySQL database connections.

As a result of the research, the stages and models were obtained (Fig. 5) according to the load effects (10).

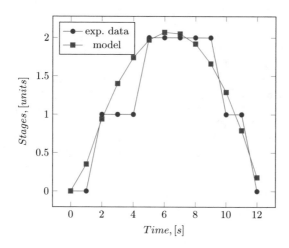

Fig. 5. Experimental data and model of stage

According to the load effect, a model of stages was determined as:

$$y = 0.06t^2 + 0.76t + 0.35;$$

The coefficient of determination was 0.91, the correlation coefficient was 0.95.

5   Conclusions

In the article proposed a model of the element as the main component of the load impacts. The input and output parameters of the element for load impacts are shown, and their characteristics are provided.

An example of load impacts using the software tool for testing open MySQL database connections is given.

Further development of the work can be directed to the study of the main criteria of stages.

References

1. Tkachenko, V., Lentz, A., Zaitsev, P., Schwartz, B., Balling, D.J., Zawodny, J.D.: High Performance MySQL, 2nd Edn. O'Reilly Media, Inc., June 2008. https://www.oreilly.com/library/view/high-performance-mysql/9780596101718/ch04.html, ISBN: 9780596101718

2. Schwartz, B.: Maatkit. Power Tools for MySQL. CPOSC (2009). https://www. xaprb.com/media/2009/10/Maatkit_CPOSC.pdf
3. Hull, S.: Even More Maatkit for MySQL. Database Journal, June 2010
4. Noach, S.: MySQL utilities for everyday use. OŘeilly MySQL Conference & Expo (2011)
5. Verbeeck, K.: Integration Services Performance Best Practices - Data Flow Optimization. SQLShack, January 2017
6. Yan, C., Cheung, A., Yang, J., Lu, S.: Understanding database performance inefficiencies in real-world web applications. In: Session 7D: Application Driven Analysis. CIKM 2017, Singapore, November 2017
7. Chen, T.-H., Shang, W., Jiang, Z.M., Hassan, A.E., Nasser, M., Flora, P.: Finding and evaluating the performance impact of redundant data access for applications that are developed using object-relational mapping frameworks. IEEE Trans. Softw. Eng. **42**(12), 1148–1161 (2016)
8. Narasimhan, B.: Database workload characteristics and their impact on storage architecture design - part 5 - Query Execution Plans, April 2015

Computer Virus Propagation Petri-Object Simulation

Inna V. Stetsenko[1(✉)] ⓘ and Vitalii Lytvynov[2] ⓘ

[1] Igor Sikorsky Kyiv Polytechnic Institute,
37 Prospect Peremogy, Kiev 03056, Ukraine
stiv.inna@gmail.com
[2] Chernihiv National University of Technology,
95 Shevchenka Street, Chernihiv 14027, Ukraine
v.vlytvynov.dept@gmail.com

Abstract. Nowadays the safety of computer networks becomes more important than other characteristics. The services which are used by a huge number of users entice hacker. Especially, online payment services are often the main target of malicious software. In the article, the computer virus SpyEye spread via Internet resources is considered. The Petri-object model of propagation process is created with the use of Petri-object simulation software written in Java. The model reproduces the events of damaging website, infecting user's resource, sending information about vulnerabilities and sending personal data. It was suggested that users are connected when they use the same website. The investigation of the spreading time for different kinds of users' connections topology is presented. It was considered not only well-known star and chain topologies but also their combinations and the topology with random connections. Simulation results confirm the virus spreading time dependence on user's connections topology.

Keywords: Simulation · Computer virus · Stochastic Petri net

1 Introduction

Usage of web services have done our life more convenient and comfortable, however, the increased popularity of these services has a negative side. Internet resource can be a source of information for a malicious person. The harm which is made by malware may be valuable.

The risk of infection for any computer resource is high nowadays through the existing vulnerabilities. Programmers can find the actual list of revealed software vulnerabilities on special resources [1] but often they cannot avoid all of them in programs. Moreover, developed software can contain vulnerability that have not been revealed yet.

Virus SpyEye considered in the paper is the botnet of third generation. Specific features of this malware and how it works and how it evolves are discussed in work [2]. The main target of this virus is online financial transactions. It gains data and information used by a person for these transactions and provides them for purposes of a

© Springer Nature Switzerland AG 2020
A. Palagin et al. (Eds.): MODS 2019, AISC 1019, pp. 103–112, 2020.
https://doi.org/10.1007/978-3-030-25741-5_11

malicious person. For example, to forge financial transaction and steal money. This virus caused $1 billion financial losses around the world and it had been listed in Top 4 malware in 2016 [3].

Effective tools for prevention malware propagation cannot be developed without knowledge of how computer virus is spread on the Internet and how fast it can be spread. Therefore, the goal of this research is to develop a model which can be used for estimation of the virus propagation process. Simulating computer virus propagation, the weak spots of network providing fast spreading of malware can be understood.

Section 1 introduce to the computer virus propagation. The second section describes related works. The next section represents the Petri-object simulation software which is used for simulation. Section 4 describes the Petri-object model of computer virus propagation which is developed and topologies which is investigated. Experimental results of simulation are presented in the next section. The time dependence on percentage of harmful sites for different topologies (chain, star, chain of stars, star of chains, random) is obtained.

2 Related Works

The spreading processes in networks have been investigated by many authors. Mathematical model of computer virus propagation in the form of differential equations was developed in [4]. The proportion of infected hosts is a variable value in these equations.

Frequently, researches are grounded on the epidemic model that assumes that a computer virus is spread as a human virus. The best review of such models and their improvement are presented in work [5]. It was analyzed by the author the effect of topology on the competing information propagation in a dynamic network.

A similar approach has been applied in work [6]. Simulation is considered there as a reproduction of a sequence of generations. The proposed model was used to compare patching and quarantine strategies of computer virus spreading prevention.

The latest researches pay great attention to the network topologies impact on malware propagation. The authors of work [7] achieved results for scale-free network topology using mathematical model. The topology was characterized by the number of nodes and the parameter of power-law exponent. In work [8] the structures of social and technological networks were investigated when they were under attack by computer virus or worm. The authors used simulation of propagation grounded on susceptible–infected–recovered epidemic model. They proposed a structural risk model.

The model of malware propagation via smart grid network is proposed in work [9]. Network simulator ns-3 was implemented [10]. The results for pandemic, endemic and contagion malware are presented.

However, the details of malware cannot be implemented by the means of epidemic model. Another way to create simulation model is Petri net. Colored Petri net is proposed to simulate the computer virus detection process in work [11]. The activities of malicious program presented by colored Petri net is stored in database. After that, the detection of the type of malicious program could be performed by comparing its scenario with those ones that have been stored.

In this research, we consider Petri-object simulation model of computer virus propagation. The significant difference from previous research is that the details of user activity and malware activity are reproduced. Website is considered as a resource that can be damaged while a user's computer resource as the one that can be infected. Site damaging is performed by a malicious person who embeds malware script. Computer resource infection will be performed by computer virus if user downloads file from a damaged site. For the first time, the model grounded on stochastic Petri net is used for the purpose of computer virus propagation simulation. In contrast to an epidemic model, the parameters of the model follow from computer virus functioning, not from assumptions.

3   Petri-Object Simulation Software

Petri-object simulation grounded on stochastic Petri net and object-oriented technology [12]. All transitions of Petri net are determined as multichannel. It means that the state of transition is described by the set of moments when tokens should do output from this transition. When tokens input is performed, transition must repeat its firing until the firing condition is completed. The input arc may be determined as informational one. The firing condition in the case of informational arc isn't changed but tokens mustn't be deleted along this arc when tokens input is performed. Such type of arc works as a permit for transition firing.

Thus, in the fragment of Petri net depicted in Fig. 1 firing condition of transition 'download file action' will be true if one token is in place 'site is harmful' and one token is in place 'uninfected'. If the firing condition is true for more than one transition the conflict between them should be resolved. It is done according to the priority and probability parameters given for each transition. After that, the tokens input is performed. The state of transition will be changed by adding new moment of tokens output and the state of its input places will be decreased by the value of appropriate arc's multiplicity. Notice that decrease will be done if only the arc is not an informational one. When the simulation time achieves a moment of tokens output, it should be performed. The state of transition will be changed by removing an appropriate value of tokens and the state of its output places will be increased on the value of appropriate arc's multiplicity.

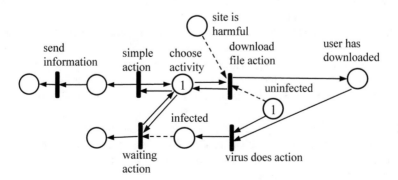

Fig. 1. The fragment of stochastic Petri net with multichannel transitions and informational arcs

Petri-object simulation software written in Java is equipped with a graphical editor to create Petri net and save it in different formats. One of them is important for creating Petri-objects. It is a saving as a method in *NetLibrary* class. The code of method is created automatically by the software in accordance to the graphical image of Petri net.

Petri-object is an object of *PetriSim* class that provides simulation according to the given stochastic Petri net. Using methods from class *NetLibrary* you can easy replicate objects with given parameters.

Petri-objects are constructive elements of Petri-object model. They can be connected in two ways: (1) to share place between two or more objects, (2) to pass token from the transition of one Petri-object to one or more places of other Petri-objects. It is proved that such Petri-objects connections guarantee that the dynamics of model will be described by the Petri net that unites nets of all Petri-objects of the model [12]. The *PetriObjModel* class provides constructing the model. It is grounded on the list of Petri-objects which have been connected. The method *go()* of this class launch simulation.

Petri-object simulation software is available on GitHub platform as an open source project PetriObjModelPaint.

4   Computer Virus Propagation Petri-Object Model

The model consists of objects 'User' and 'Malice' which simulate the behavior of person using sites and person embedding malware correspondently. The 'User' reproduces events such as to choose a site, to do a simple action, to download a file. It has states 'infected' and 'uninfected'. Simple action such as reading information cannot be harmful but the action of downloading a file is used by a malware to make user's resource infected. When the user is infected the events 'send information', 'waiting action' and 'download file' become permitted. User's computer resource becomes an information and data source of malware that provides knowledge about sites vulnerabilities user have attended, and about data user have loaded. Information about site vulnerabilities exploits by malware for virus propagation. Stored data can be used for stealing money.

The behavior of the user who interacts with one site presents by Petri-object 'Website'. One user can exploit a few sites, so object 'User' aggregates a list of sites. The places 'user', 'infected', 'uninfected' are share for all sites of this list. The net of Petri-object 'Website' is depicted in Fig. 2.

The 'Malice' object is created as Petri-object that reproduces malware events such as to insert malice script, to send spam, to initiate theft (Fig. 3). The virus is propagated by embedding malice script in website code if it has suitable vulnerability. If information about vulnerability has been received the event of the site being damaged is permitted. It is suggested that malice person is able to interact with all users which are added to the model.

The fact that users can apply the same sites is represented in model by setting share sites for different pairs of users. It can be done by sharing places 'sites', 'harmful', and 'unharmed' of corresponding 'Website' objects. In the Fig. 4 the example of the 'User' objects connection is presented.

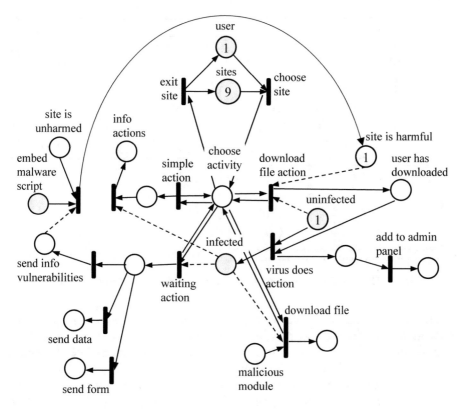

Fig. 2. The net of Petri-object 'Website'

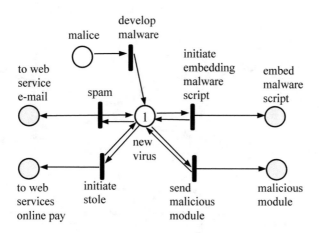

Fig. 3. The net of Petri-object 'Malice'

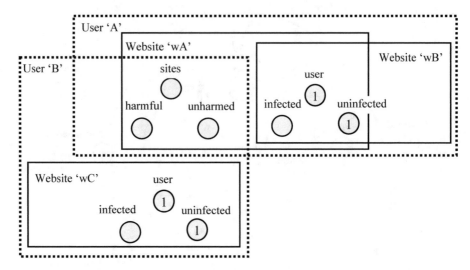

Fig. 4. The 'User' objects connection: site 'wA' is shared for users 'A' and 'B'

The users' connections create a topology. To investigate the impact of topology on malware propagation we consider different kinds of them (Fig. 5). Chain structure is formed when each user has only two connections with other users. Star structure is deduced when one of the users are connected with all other users. Combining these types of topology, we obtain the chain of stars structure and the star of chains structure. Also, random pairs of connections generated with given upper limit of the number of connections can create a structure which we called 'chaos' (Fig. 6). This last type of structure should consider as the closest to real life. Especially, it is essential for modern conditions when the structure of users' connections tends to change frequently.

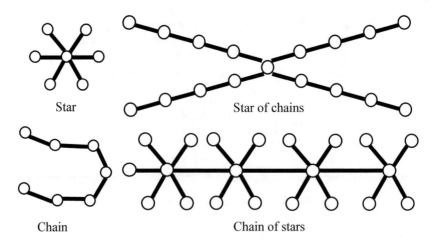

Fig. 5. The types of topologies of users' connections: chain, star, chain of stars, star of chains

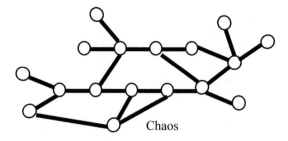

Chaos

Fig. 6. The topology with random users' connections (the upper limit of connections is equal 4).

Aggregating 'User' objects connected to each other according to the given topology and the 'Malice' object connected to all users we obtain the model of virus propagation. Experimental investigation of the model has allowed to achieve results presented in the next section.

5   Simulation Results

Simulation results are obtained for the model which has at start moment only one harmful site. It causes the user's computer resource infecting in the case of user have chosen a download action. After that, malice person can receive information about sites' vulnerabilities and becomes able to insert malice script into an unharmed site. In this way virus is propagated. The values that are measured as a result of simulation are the moments when the sites become harmful. Then the percentage of harmful sites for every moment can be calculated.

We will suggest that the users are very cautious and download the file proposed by the website in 1 case of 2000. Changing the type of topology, we can see that the chain structure shows the slowest propagation whereas the star structure shows the fastest. The simulation results in the case of 51 users one of which is infected at start moment are presented in Fig. 7. The number of sites is equal to $51 \cdot 4 - 50 = 154$ for every structure.

Thus, Internet resources with 'chain' topology of connections can be damaged in the biggest time. But this type of structure is the least possible in real life. A closer topology to reality is 'chain of stars' which is showed the next result after the best. After model verification on simple topologies more complicated random structures can be considered. We propose to create an arbitrary structure by generating random connections between users with the limited maximum number of connections one user has and to call it 'chaos' topology. The results in the case of upper limit of connections equals 8 are depicted in Fig. 8. Despite the fact that the structures were generated by the same algorithm they show different time propagation. Hence, the topology of connections has a strong impact even if the number of connections is the same. Large enough number of maximum connections provides that all sites are damaged during time simulation. But in the case of maximal number equals 4 we observed the random generated disconnected structure which, evidently, cannot achieve the state when all sites are damaged.

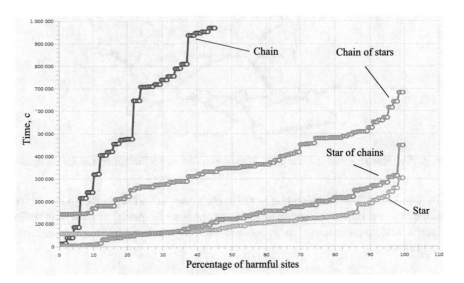

Fig. 7. The time dependence on percentage of harmful sites for different topologies (chain, star, chain of stars, star of chains).

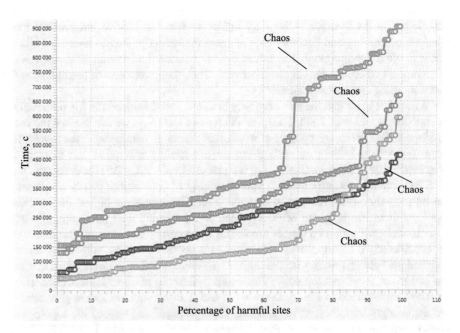

Fig. 8. The time dependence on percentage of harmful sites for 'chaos' topologies (all structures are different and generated random, the upper limit of connections is equal 8).

Another experiment shows the impact of the chance of downloading file action (Fig. 9). Chain structure has been removed in this graphic because it has the time much greater than other structures. It is observed that the time of all sites damage is rapidly decreased when the chance of downloading file action is increased. The difference between topologies will be significant if the value of the chance is small.

Thus, experimental results show the model availability to reproduce propagation of computer virus via Internet resources. We obtain strong dependence of the percentage of harmful sites on the topology of users' connections. According to the results, the structure that contains chain fragments is highly recommended.

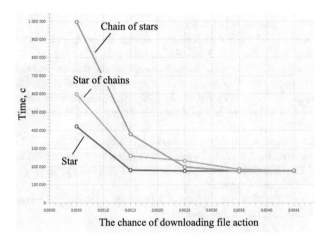

Fig. 9. The time dependence on chance of downloading file action for different topologies.

6   Conclusions

The main result we obtained is that computer virus propagation via Internet resources depends on users' connections topology. Chain fragments included in the topology significantly increase the time of virus propagation. Simulation using Petri-object model allows to take into account specific features of malware. Parameters of the model are formulated as characteristics of malware behavior whereas epidemic models or mathematical models require parameters assumptions. The experimental results confirm the model availability to evaluate the impact of users' connections structure.

Acknowledgment. This work is funded by the NATO SPS Project CyRADARS (Cyber Rapid Analysis for Defense Awareness of Real-time Situation), Project SPS G5286.

References

1. National Technology Laboratory: National Vulnerability Database (NVD). NVD Data Feeds. https://nvd.nist.gov/vuln/data-feeds. Accessed 20 April 2019

2. Sood, A.K., Enbody, R.J., Bansal, R.: Dissecting spyeye – understanding the design of third generation botnets. Comput. Netw. **57**(2), 436–450 (2013)
3. iZOOlogic: Top 4 Malware – Financial Trojans – Zeus, Carberp, Citadel and SpyEye (2016). https://www.izoologic.com/2016/10/15/top-4-malware-financial-trojans-zeus-carberp-citadel-and-spyeye/. Accessed 20 April 2019
4. Serazzi, G., Zanero, S.: Computer virus propagation models. In: Calzarossa, M.C., Gelenbe, E. (eds.) Performance Tools and Applications to Networked Systems. Lecture Notes in Computer Science, vol. 2965, pp. 26–50. Springer, Heildelberg (2004)
5. Valler, N.C.: Spreading Processes on Networks: Theory and Applications. University of California Riverside (2012) https://escholarship.org/uc/item/6r76d0rg. Accessed 19 April 2019
6. Asllani, A., Ali, A.: Using simulation to investigate virus propagation in computer networks. Netw. Commun. Technol. **1**(2), 76–85 (2012)
7. Yang, L., Yang, X.: The effect of network topology on the spread of computer viruses: a modeling study. Int. J. Comput. Math. **94**(8), 1–19 (2017)
8. Guo, H., Cheng, H.K., Kelley, K.: Impact of network structure on malware propagation: a growth curve perspective. J. Manage. Inform. Syst. **33**(1), 296–325 (2016)
9. Eder-Neuhauser, P., Zseby, T., Fabini, J.: Malware propagation in smart grid networks: metrics, simulation and comparison of three malware types. J. Comput. Virol. Hacking Tech. **14**(3), 1–17 (2018)
10. ns-3 Network Simulator. https://www.nsnam.org/. Accessed 25 April 2019
11. Jasiul, B., Szpyrka, M., Śliwa, J.: Detection and modeling of cyber attacks with Petri nets. Entropy **16**(12), 6602–6623 (2014). Entropy-Based Applied Cryptography and Enhanced Security for Ubiquitous Computing
12. Stetsenko, I.V.: Theoretical foundations of Petri-object modeling of systems. Math. Mach. Syst. **4**, 136–148 (2011). (In Russian)

Retransmission Request Method for Modern Mobile Networks

Zaitsev Sergei[1(✉)], Vasylenko Vladyslav[2], Trofymchuk Oleksandr[2], and Tkach Yuliia[1]

[1] Chernihiv National University of Technology,
95 Shevchenka Street, Chernihiv 14035, Ukraine
serzal979@gmail.com, tkachym79@gmail.com
[2] Institute of Telecommunications and Global Information Space of NAS
of Ukraine, 54 Volodymyrska Street, Kyiv 01030, Ukraine
vladvasilenko9@gmail.com, itelua@kv.ukrtel.net

Abstract. The article deals with the issue of improving the functioning efficiency of LTE and WiMAX radio networks, built according to the scheme with a hybrid automatic request for retransmission. The article describes a modified method of a hybrid automatic request formation for retransmission in the conditions of uncertainty arising from the impact of powerful interference in the data transmission channel. The method is based on the retransmission of data bits, which were determined to be erroneous after decoding, using additional information about the logarithmic likelihood functions (LLF) for these bits when calculating the resulting likelihood functions by the turbo decoder based on previous retransmissions. In contrast to the known results, in case of the need to re-request the transfer, not the entire data block is transmitted, but only those bits that are defined as erroneous. The results of the information transmission system simulation showed that the use of a modified method of a hybrid automatic request formation for retransmission made it possible to reduce the number of retransmitted data bits by $13 \sim 16$ times.

Keywords: Turbo codes · Automatic retransmission request · Adaptation · Uncertainty · Credibility functions

1 Target Settings

Reliability of data transmission is one of the main problems in the field of wireless communications. To improve the reliability of data transmission, automatic repeat request systems for retransmission (Automatic Repeat reQuest - ARQ) were introduced. There are several ARQ methods: Stop-and-wait ARQ, Go-Back-N ARQ, ARQ selective retry (selective failure). These methods are widely applied in various protocols, standards of data transmission systems. For example, in the TSR reliable data transfer protocol, the Go-Back-N ARQ method is used. However, the application of ARQ methods leads to a reduction in data throughput. To solve the problem of reducing bandwidth, ARQ methods are used in conjunction with Forward Error Correction (FEC) methods to create a hybrid automatic request for retransmission (HARQ - Hybrid

© Springer Nature Switzerland AG 2020
A. Palagin et al. (Eds.): MODS 2019, AISC 1019, pp. 113–121, 2020.
https://doi.org/10.1007/978-3-030-25741-5_12

ARQ) systems [1–7]. The role of FEC is to reduce the frequency of data packets retransmissions by correcting errors in the received packets, that leads to an increase in the throughput of data transmission channels.

2 Formulating the Goals of the Article

The purpose of the article is to develop a modified method of a hybrid automatic request formation for re-transmission to improve the reliability of information transmission in conditions of high noise levels in the data transmission channel. The method is based on the retransmission of data bits, which were determined to be erroneous, using the LLF additional information for these bits.

3 Actual Scientific Researches and Issues Analysis

The essence of the method lies in the re-transmission of data bits, which were determined to be erroneous, using additional information about LLF for these bits when calculating the resulting likelihood functions by the turbo code decoder. Existing methods of a hybrid automatic request formation for retransmission use an entire block of data with each subsequent transfer, while significantly reducing the throughput of data transmission channels.

The high efficiency of turbo codes largely depends on the principles of the code combinations formation and the probabilistic decoding algorithms developed for them, which take into account both a posteriori and a priori information to improve the reliability of decoding. The correcting ability of the code is based on performing several decoding steps or decoding iterations. This property was used as the basis for the construction of decoding algorithms for turbo codes in order to increase the decoding reliability when the a posteriori information of the turbo code decoder after the interleaving or deinterleaving operation is used as a priori information for the next decoder.

Figure 1 shows a block diagram of a turbo code decoder, which consists of two elementary decoders, each of which decodes information generated by the corresponding component recursive systematic convolutional code (RSCC), as well as two interleavers and two deinterleavers. The interleavers are similar to those used in the encoder.

Each turbo code decoder computes logarithmic likelihood functions of the transmitted information bits, represented as "soft" solutions. In this case, the modulus of the value obtained is proportional to the plausibility (reliability) of the transmitted bit, and the sign corresponds to the symbol value: minus – to zero, and plus – to one.

The logarithmic relations of the likelihood functions for each transmitted bit, obtained in the decoding process by the component decoder of the turbo code, are rearranged or departed before each subsequent decoder (depending on the decoder used), which leads to a decrease in the correlation links between information and check symbols, as well as to an increase in corrective turbo code properties.

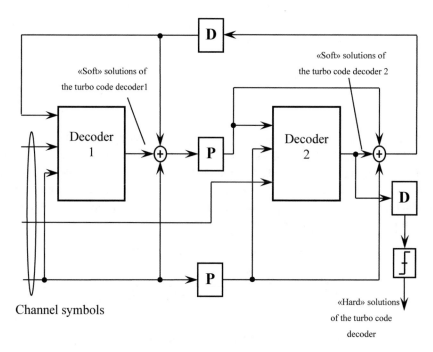

Fig. 1. Block diagram of a turbo code decoder

In the communication channel, the transmitted symbols are affected by noise and intentional interference, which can be represented as a band-limited additive white Gaussian noise. As a result, the systematic and test symbols at the channel output (a discrete-continuous channel is considered) will be random variables distributed according to the normal law: $y_t^C = x_t^C + n_t^*$, $y_t^{\Pi 1} = x_t^{\Pi 1} + n_t^{**}$, $y_t^{\Pi 2} = x_t^{\Pi 2} + n_t^{***}$, where $n_t^*, n_t^{**}, n_t^{***}, t \in \overline{1, N}$ – are samples of white Gaussian noise. From the channel output, the systematic and check symbols y_t^C, $y_t^{\Pi 1}$, $y_t^{\Pi 2}$, $t \in \overline{1, N}$ enter the input of the turbo code decoder.

Turbo codes are decoded using the decoding algorithm for the maximum a posteriori probability MAP (maximum a posteriori probabilities), which calculates the a posteriori probability of each decoded symbol, minimizing the error probability of the information symbol (bit) [16].

Decoding takes place in two directions: direct and transient recursions are calculated for each state of the turbo code in the first direction (from the beginning of the block to the end), reverse recursions are calculated in the second direction (from the end of the block to the beginning) using transient recursions obtained in the first direction calculations.

When requesting a retransmission, the RSCC polynomials are modified, the corresponding trellis state diagram is formed, the corresponding recursion and likelihood functions are calculated. Consider these calculations in details.

The logarithmic ratio of the likelihood functions $L(u_k)$ for the transmitted random binary variable u_k is defined as follows:

$$L(u_t) \underline{\underline{\Delta}} \log\left(\frac{P(u_t = 1/y_t)}{P(u_t = 0/y_t)}\right). \tag{1}$$

The decision on the decoding results can be made by the sign, $L(u_t)$, i.e.

$$\tilde{u}_t = \text{sign}[L(u_t)]. \tag{2}$$

The logarithmic ratio of the likelihood functions of the transmitted bit $L(u_t)$ depends on the channel information $L_c(y_t)$, a priori information about the transmitted bit $L_a(x_t)$ and the a posteriori LLF, produced directly by the decoder itself $L_e(x_t)$. Therefore, when decoding a bit y_t for computations by the first decoder at the decoding iteration $j, j \in \overline{1, I}$, where I – is the total number of decoding iterations, the expression (1) can be rewritten in the following way [15]:

$$L^{1,j}(x_t) = \log \frac{\sum\limits_{\substack{(s',s) \\ u_t=1}} \tilde{\alpha}_{t-1}^{(1)}(s') \cdot \tilde{\beta}_t^{(1)}(s) \cdot \gamma_t^{(1)}(s',s)}{\sum\limits_{\substack{(s',s) \\ u_t=0}} \tilde{\alpha}_{t-1}^{(1)}(s') \cdot \tilde{\beta}_t^{(1)}(s) \cdot \gamma_t^{(1)}(s',s)} \tag{3}$$

$$= L_c^{1,j}(y_t) + L_a^{1,j}(x_t) + L_e^{1,j}(x_t)$$

where $L_c^{1,j}(y_t)$ – is the channel information, $L_e^{1,i}(x_t)$ – a posteriori LLF data bits x_t.

Accordingly, for the second decoder we get:

$$L^{2,j}(x_t) = \log \frac{\sum\limits_{\substack{(s',s) \\ u_t=1}} \tilde{\alpha}_{t-1}^{(2)}(s') \cdot \tilde{\beta}_t^{(2)}(s) \cdot \gamma_t^{(2)}(s',s)}{\sum\limits_{\substack{(s',s) \\ u_t=0}} \tilde{\alpha}_{t-1}^{(2)}(s') \cdot \tilde{\beta}_t^{(2)}(s) \cdot \gamma_t^{(2)}(s',s)} \tag{4}$$

$$= L_c^{2,j}(y_t) + L_a^{2,j}(x_t) + L_e^{2,j}(x_t)$$

Next, the posteriori LLF data bits x_t, produced by the decoder itself are calculated, $- L_e^{1,j}(x_t)$:

$$L_e^{1,j}(x_t) = L^{1,j}(x_t) - L_c^{1,j}(y_t) - L_a^{1,j}(x_t). \tag{5}$$

After the interleaver P, the a posteriori LLF $L_e^{1,i}(x_t)$ is converted to the a priori LLF $L_a^{2,j}(x_t)$: $L_a^{2,j}(x_t) = f_1\left(L_e^{1,j}(x_t)\right)$, where $f_1(\cdot)$ – is the function performing the interleaving operations, and is fed to decoder 2. Decoder 2 performs similar calculations to obtain the value $L_e^{2,i}(x_t)$:

$$L_e^{2,j}(x_t) = L^{2,j}(x_t) - L_c^{2,j}(y_t) - L_a^{2,j}(x_t). \tag{6}$$

After executing the deinterleaving operation D: $L_a^{1,j+1}(x_t) = f_2(L_e^{2,j}(x_t))$, where $f_2(\cdot)$ – is the function, performing the deinterleaving operations, the value $L_a^{1,j+1}(x_t)$ is used as a priori for decoder 1 iteration $j + 1$. After all iterations of decoding «hard» decisions about the transmitted bit are completed: $\tilde{u}_t = \text{sign}[L(u_t)]$.

Figure 2 shows a block diagram of a modified adaptive iterative decoder TK, taking into account the LLF, obtained in the previous requests for retransmission.

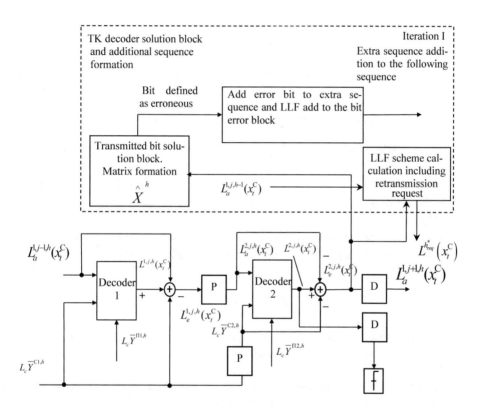

Fig. 2. Block diagram of the modified adaptive iterative decoder TK

Algorithm for generating a hybrid automatic request for retransmission.
Step 1. Input the source data:

– number of automatic retransmission requests h, $h \in \overline{1, H}$;

– parameters of the turbo code encoder $\{I, N, P, K, \overrightarrow{G}, R\}$, where I – is the number of iterations of decoding the turbo code, N – is the data block size in bits, P – is a type of interleaver, K – is the number of component encoders (decoders), $\overrightarrow{G}^H = (g_1^0, g_0^0, g_1^1, g_0^1, \ldots, g_1^H, g_0^H,)$ – is a vector of turbo code encoder polynomials, R – is the encoding speed of the turbo code.

Step 2. The automatic retransmission request parameter $h = 1$, is needed to track the number of requests.

Step 3. Calculations LLF data bits x_t, $t \in \overline{1,N}$ i-th decoder, $i \in \overline{1,2}$, j-th decoding iteration, $j \in \overline{1,I}$, for all bits of block length N, decoder 1 and 2, decoding iterations $j \in \overline{1,I}$, where I – is the total number of decoding iterations.

$$
\begin{aligned}
L^{i,j,h}(x_t) &= \log \frac{\displaystyle\sum_{\substack{(s',s) \\ u_t=1}} \tilde{\alpha}_{t-1}^{(i)}(s') \cdot \tilde{\beta}_t^{(i)}(s) \cdot \gamma_t^{(i)}(s',s)}{\displaystyle\sum_{\substack{(s',s) \\ u_t=0}} \tilde{\alpha}_{t-1}^{(i)}(s') \cdot \tilde{\beta}_t^{(i)}(s) \cdot \gamma_t^{(i)}(s',s)} \\[2mm]
&= \log \frac{\displaystyle\sum_{\substack{(s',s) \\ u_t=1}} \tilde{\alpha}_{t-1}^{(i)}(s') \cdot \tilde{\beta}_t^{(i)}(s) \cdot \exp\left[\frac{1}{2} \cdot \left(x_t^{C,h} \cdot \left(L_a^{i,j,h}(x_t^{C,h}) + L_c \cdot y_t^{C,h}\right) + L_c \cdot y_t^{\Pi i,h} x_t^{\Pi i,h}\right)\right]}{\displaystyle\sum_{\substack{(s',s) \\ u_t=0}} \tilde{\alpha}_{t-1}^{(i)}(s') \cdot \tilde{\beta}_t^{(i)}(s) \cdot \exp\left[\frac{1}{2} \cdot \left(x_t^{C,h} \cdot \left(L_a^{i,j,h}(x_t^{C,h}) + L_c \cdot y_t^{C,h}\right) + L_c \cdot y_t^{\Pi i,h} x_t^{\Pi i,h}\right)\right]} \quad (7) \\[2mm]
&= L_c^{i,j,h}(y_t) + L_a^{i,j,h}(x_t) + L_e^{i,j,h}(x_t).
\end{aligned}
$$

Formation of LLF values matrixes on transmitted bits x_t, $t \in \overline{1,N}$, block of size N for the i-th decoder, $i \in \overline{1,2}$, j-th decoding iteration, $j \in \overline{1,I}$: $L^h = [L^{i,j,h}(x_1) \, L^{i,j,h}(x_2) \ldots L^{i,j,h}(x_N)]$.

Step 4. Calculation of the a posteriori LLF data bit x_t, $t \in \overline{1,N}$ i-th decoder, $i \in \overline{1,2}$, j-th decoding iteration, $j \in \overline{1,I}$, for all block bits of length N, decoder 1 and 2, decoding iterations $j \in \overline{1,I}$:

$$
L_e^{i,j,h}(x_t) = L^{i,j,h}(x_t) - L_c^{i,j,h}(y_t) - L_a^{i,j,h}(x_t). \quad (8)
$$

Values matrixes formation of a posteriori LLF about transmitted bits x_t, $t \in \overline{1,N}$, block of size N for the i-th decoder, $i \in \overline{1,2}$, j-th decoding iteration, $j \in \overline{1,I}$: $L_e^h = [L_e^{i,j,h}(x_1) \, L_e^{i,j,h}(x_2) \ldots L_e^{i,j,h}(x_N)]$.

Step 5. After the completion of all decoding iterations, «hard» estimates of the decoded bits are made. If $h = H$, then go to step 12.

Step 6. Decision on the value of the decoded information bits:

$$
\hat{x}_t^h = \begin{cases} 1, & L_e^{i,j,h}(x_t) > 0 \\ 0, & L_e^{i,j,h}(x_t) < 0 \end{cases}.
$$

Formation of decoded data bits matrix $\hat{X}^h = [\hat{x}_1^h \, \hat{x}_2^h \ldots \hat{x}_N^h]$.

Errors presence control in the received data block. The definition of the matrix elements $\hat{X}^h$ bits which are decoded as erroneous. Formation, respectively, of the matrix X_{err}^h, which contains erroneously decoded data bits: $X_{err}^h = [x_{1,pos} \, x_{2,pos} \ldots x_{k,pos}]$, where pos – is the bit position in the block.

Step 7. Formation of a HARQ signal, which is transmitted to the decoder to modify the decoding algorithm, and is fed through a feedback channel to retransmit a block X_{err}^h of size K, which contains erroneous data bits, along with the next block.

Шаг 8. Parameter for automatic retransmission requests $h = h + 1$. If $h < H$, then go to step 9, if not – go to step 5.

Step 9. Formation of the next data block $X_{err} = [\hat{X}_{err}^h, \hat{X}_{err}^{h-1}, \ldots, \hat{X}_{err}^1]$.

Step 10. Performing the basic steps of coding, transferring data bits to a discrete-continuous channel, decoding a received data block. Calculations LLF data bits x_m,

$m \in \overline{1, N + \sum_{r=1}^{h-1} K_r}$, where K_r – is the number of error bits for the corresponding block.

Formation of LLF matrixes about transmitted bits x_m, $m \in \overline{1, N + \sum_{r=1}^{h-1} K_r}$ for a block of

size $(N + \sum_{r=1}^{h-1} K_r)$ for the i-th decoder, $i \in \overline{1, 2}$, j-th decoding iteration, $j \in \overline{1, I}$:

$L^h = [L_F^h, L_{S\,err}^h] = [L_F^h, L_{err}^{h-1}, \ldots, L_{err}^1]$, where L_F^h – is the matrix of the LLF transmitted

data block of N size, $L_{S\,err}^h$ – is the matrix of erroneous data bits of size $\sum_{r=1}^{h-1} K_r$;

$$L_F^h = [L_F^{i,j,h}(x_1)\,L_F^{i,j,h}(x_2)\ldots L_F^{i,j,h}(x_N)],$$
$$L_{err}^{h-1} = [L_S^{i,j,h-1}(x_1)\,L_S^{i,j,h-1}(x_2)\ldots L_S^{i,j,h-1}(x_{K_r})], \ldots,$$
$$L_{err}^1 = [L_S^{i,j,1}(x_1)\,L_S^{i,j,1}(x_2)\ldots L_S^{i,j,1}(x_{K_r})].$$

Step 11. Performing operations for calculating the total LLF of the transmitted bits, which are classified as erroneous:

$$L_{err}^{h-1*} = [L_S^{i,j,h-1}(x_1) + L_S^{i,j,h-2}(x_1)\,L_S^{i,j,h-1}(x_2) + L_S^{i,j,h-2}(x_2)\ldots L_S^{i,j,h-1}(x_{K_r}) + L_S^{i,j,h-2}(x_{K_r})].$$

Go to step 5.

Step 12. Transfer the decoded block to the data receiver.

Evaluation of the information transfer reliability characteristics using the proposed method of a hybrid automatic request formation for retransmission is carried out using the method of simulation. To compare the proposed results, the fourth-generation 4G LTE-Advanced mobile communication standard was chosen as an analogue.

Figure 3 shows a plot of the normalized value $Q = S/(n * N)$ (S – is the number of bits transmitted during the retransmission request, n – is the number of blocks, N – is the number of bits in the block) versus signal-to-noise ratio in the channel E_b/N_J without using and applying the proposed method.

When modeling, the turbo code with two component encoders, polynomial generators (1,23/21), S-random interleaver (deinterleaver), number of bits in the transmitted block $N = 6144$, *Log Map* decoding algorithm, TK coding rate $R = 1/3$, 8 decoding iterations is used. Analysis of the simulation results shows that using the modified method of a hybrid retransmission request formation, the number of retransmitted data bits has decreased by 13 $\sim$ 16 times.

Q

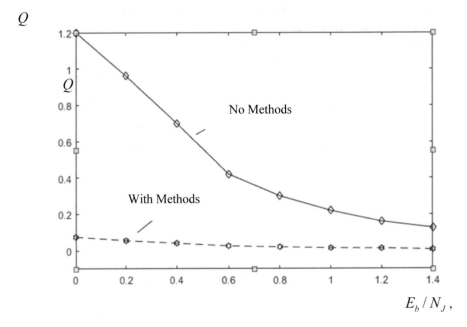

Fig. 3. The results of the transmission characteristics of additional bits simulation for a block of data size $N = 6144$

4 Conclusions

1. The article proposes a modified method of a hybrid automatic re-transmission request formation to improve the reliability of information transmission in conditions of high noise levels in the data transmission channel.
2. The method is based on the retransmission of data bits, which were determined to be erroneous, using additional information about the logarithmic ratios of the likelihood functions for these bits when calculating the resulting likelihood functions by the turbo code decoder. In contrast to the known results, in case of the need to re-request the transfer, not the entire data block is transmitted, but only those bits that are defined as erroneous.
3. The results of the information transmission system simulation show that the use of a modified method of a hybrid automatic re-transmission request formation make it possible to reduce the number of retransmitted data bits by 13 ~ 16 times.

References

1. Abudakar, B.: Automatic repeat request (Arq) protocols. Int. J. Eng. Sci. **6**(5), 64–66 (2017). https://doi.org/10.9790/1813-0605016466
2. Saber, H., Marsland, I.: An incremental redundancy hybrid ARQ scheme via puncturing and extending of polar codes. IEEE Trans. Commun. **63**, 3964–3973 (2015). https://doi.org/10.1109/tcomm.2015.2477082

3. Cheolsu, H., Seok, H.K.: EMI reduction algorithm using enhanced-HARQ implementation for controller area network. Int. J. Appl. Eng. Res. **12**(21), 11124–11129 (2017)
4. Santosh, C., Swarna, J.: Review on cross layer hybrid scheme in WMSNS using error correcting codes for energy efficiency. Int. J. Adv. Res. Innov. Ideas Educ. **2**(5), 315–322 (2017)
5. Yong, J., Peng, R.: Adaptive cooperative FEC based on combination of network coding and channel coding for wireless sensor networks. J. Netw. **9**(2), 481–487 (2014)
6. Simmi, G., Anuj, K., Tyagi, A.: An introduction to various error detection and correction schemes used in communication. Int. J. Appl. Res. **2**(8), 216–218 (2016)
7. Farouq, M., Yahya, O., Ismail, K., Adel, B.: Maximizing throughput of SW ARQ with network coding through forward error correction. Int. J. Adv. Comput. Sci. Appl. **6**(6), 291–297 (2015)
8. Sachin, K., Nikhil, S., Nisarga, C., Apeksha, S., Usha, S.: A review of hybrid ARQ in 4G LTE. Int. J. Adv. Res. Innov. Ideas Educ. **1**(3), 160–165 (2015)
9. Zaitsev, S.V., Kazymyr, V.V.: Method of adaptive decoding in case of information transmission in condition of influence of deliberate noise. Radioelectron. Commun. Syst. **58**, 30–40 (2015). https://doi.org/10.3103/s0735272715050039
10. Zaitsev, S.V.: Structural adaptation of the turbo code coder and decoder for generating the transmission repeat request under conditions of uncertainty. Radioelectron. Commun. Syst. **60**, 18–27 (2017). https://doi.org/10.3103/s0735272717010034
11. Zaslavsky, A., et al.: Self-organizing intelligent network of smart electrical heating devices as an alternative to traditional ways of heating. In: Proceedings of SPIE, Photonics Applications in Astronomy, Communications, Industry and High Energy Physics Experiments 2017, Poland, vol. 10445, p. 104456K, 7 August 2017. https://doi.org/10.1117/12.2281225
12. Lin, S., Costello, D., Miller, M.: Automatic-repeat-request error control schemes. IEEE Commun. Mag. **22**, 5–17 (1984)
13. Cao, L., Shi, T.: Turbo codes based hybrid ARQ with segment selective repeat. Electron. Lett. **40**(18), 1140–1141 (2004)
14. Chase, D.: Code combining – a maximum-likelihood decoding approach for combining an arbitrary number of noisy packets. IEEE Trans. Commun. **33**(5), 385–393 (1985)
15. Hollandm, I., Zepernick, H., Caldera, M.: Soft combining for hybrid ARQ. IET Electron. Lett. **41**(22), 1230–1231 (2005)
16. Woodard, J., Hanzo, L.: Comparative study of turbo decoding techniques: an overview. IEEE Trans. Veh. Technol. **49**(6), 2208–2232 (2000)

Method of Deformed Stars for Multi-extremal Optimization. One- and Two-Dimensional Cases

Vitaliy Snytyuk$^{(\boxtimes)}$ (iD)

Taras Shevchenko National University of Kyiv, Kyiv 04116, Ukraine
snytyuk@gmail.com

Abstract. This paper describes the task of optimizing a multi-extremal function, which in general can be given analytically, tabularly or algorithmically. The method of deformed stars is developed, which belongs to the class of evolutionary methods and allows to take into account the relief of the investigated function. Its advantages are the speed of convergence and result accuracy in comparison with other evolutionary methods. The obtained results of experiments allow us to conclude that the proposed method is applicable to solving problems of finding optimal (suboptimal) values, including non-differentiated functions.

Keywords: Function · Optimization · Deformed stars method

1 Introduction

The problem of continuous and discrete optimization of functional dependencies is well known as the methods of its solution. Traditionally, two approaches are used: the first is the classical integro-differential approach, the technologies of the second approach are called stochastic optimization. In the first case, the function whose optimal value is sought is subject to tight constraints, the solution of the problem using the second approach does not guarantee finding a global optimum and requires significant computing resources.

The features above and new problems that have emerged in recent years, as well as the fact that functional dependencies can be set not only analytically, but also tabularly or algorithmically, are undifferentiated and multi-extremal, triggered the emergence and relevance of the use of new methods and algorithms that attributed to the evolutionary paradigm of simulation. Important features of such methods are imitation of the technology of natural systems functioning, inspired by natural evolution, and their adaptation to environmental conditions. As a rule, evolutionary methods are population and implement a strategy of parallel stochastic search using special operators.

Researches divide the approaches that are implemented in evolutionary methods into two classes. In methods belonging to the first class, the binary coding of potential solutions is used, which is an approximate analogy of the DNA structure. New solutions are obtained by crossing parental genotypes, which is an analogue of sexual reproduction. In methods of the second class, the process of obtaining new potential

A. Palagin et al. (Eds.): MODS 2019, AISC 1019, pp. 122–130, 2020.
https://doi.org/10.1007/978-3-030-25741-5_13

solutions resembles asexual reproduction. In this case, as a rule, the offsprings are placed from parents at a distance, which is distributed under the normal law.

Problems in applying methods of both the first and second classes are the low convergence rate and access to local optimums if the target function is multi-extremal.

2 Brief Analysis of Research, Publication, and State-of-the Art

It is known that the main evolutionary paradigms appeared in the second half of the 20th century. The appearance of the term "genetic algorithms" is associated with the printing of J. Holland's monograph "Adaption in Natural and Artificial Systems" in 1975, and the method of "evolutionary strategies" was developed by German students Rechenberg and Schwefel in 1962–1964 [1]. These methods have been improved by many scientists, they were used to creating hybrid technologies for solving optimization problems of practice. In 1997 was proved a theorem that there is no single method that would allow to find an optimal solutions to all optimization problems [2].

Therefore, the development of new methods, based on the principles of natural evolution and recommended for use in solving certain classes of problems was continued. Thus, the methods of Differential Evolution [3], Harmony Search [4], Cuckoo Search [5], Symbiotic Organisms Search [6] and many others were proposed. Along with the development of methods that were inspired by the behavior of certain objects of nature, hybrid methods were developed, in which several "mono-methods" were integrated [7]. Methods with vertical integration of components were called memetic [8], methods with horizontal interaction are called cooperative and coevolution [9].

In recent years are being developed multi-objective optimization methods based on the evolutionary paradigm [9, 10] and methods for solving problems of data mining, in particular, classification, clustering [11], allocation of associations and sequences, incl. and with the use of neural network technologies.

Note that all these methods have universal nature and designed to optimize any objective function. At the same time, taking into account the relief of the objective function can significantly accelerate the convergence of one or another method.

3 Problem Formulation

Consider the problem (one-dimensional case)

$$f(x) \rightarrow max, x \in [a, b], a, b \in R. \tag{1}$$

About characteristics of the function $f(x)$ nothing is known. As already mentioned earlier, the function can be given tabularly, and then we can get its model in the form of a neural network, or algorithmically.

In the two-dimensional case, we write the problem as:

$$f(x_1, x_2) \rightarrow max, x_1 \in [p_1, p_2], x_2 \in [q_1, q_2], p_1, p_2, q_1, q_2 \in R. \tag{2}$$

As an example we can see a multi-extremal function (see Fig. 1)

$$f(x) = (x^2 - 2x + 1) \sin(10x)/x \tag{3}$$

and the Schwefel's function (see Fig. 2)

$$f(x_1, x_2) = -\left(x_1 sin \sqrt{|x_1|} + x_2 sin \sqrt{|x_2|}\right). \tag{4}$$

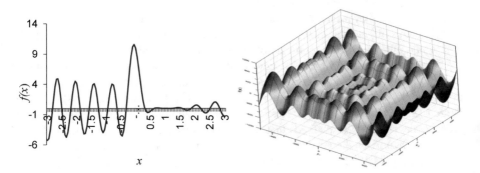

Fig. 1. $f(x) = (x^2 - 2x + 1) \sin(10x)/x$ **Fig. 2.** Schwefel's function

4   Method of Deformed Stars. One-Dimensional Case

Assume that we need to solve a problem of type (1). The method of deformed stars has such a sequence of steps.

Step 1. Let t = 0 (iteration number).

Step 2. Generate the initial population of potential solutions.
$P_t = \{x_1, x_2, \ldots, x_n\}$, uniformly distributed in the interval (a, b), $|P_t| = n$.

Step 3. For all $x_j \in P_t$ we find $f_j = f(x_j)$, that form the population $F_t = \{f_j\}_{j=1}^n$.

Step 4. Generate elements of the new population P_z as:

$$x_j^z = x_j + \xi(N(0, \sigma_j)), j = \overline{1, n},$$

where $\xi(*)$ is a randomly distributed random variable. If $x_j^z < a$, then $x_j^z = x_j^z + (b - a)$; if $x_j^z > b$, then $x_j^z = x_j^z + (a - b)$. For all $x_j^z \in P_z$ we find $f_j^z = f(x_j^z)$, that form the population $F_z = \{f_j^z\}_{j=1}^n$.

Step 5. Elements of the third population $P_s = \{x_k^s\}_{k=1}^n$ are obtained as follows:

$$x_k^s = (x_i + x_j)/2, \text{ where } i, j = random\{1, 2, \ldots, n\}, k = \overline{1, n}.$$

Similarly, for all $x_k^s \in P_s$ we find $f_k^s = f(x_k^s)$, that form the population

$$F_s = \{f_k^s\}_{k=1}^n.$$

Step 6. To aggregate the elements of the populations we obtain new sets $P_t \cup P_z \cup P_s = P$ and $F_t \cup F_z \cup F_s = F$. Elements of population F are sorted by descending. $t = t + 1$.

Step 7. We create a new population P_t from elements of the population P that correspond to the first n elements of population F.

Step 8. If the stop condition is not fulfilled, the iterative process continues (go to step 3). If the stop condition is satisfied, then the value of the potential solution, which corresponds to the maximum value of the function f, will be the solution of the problem (1).

The conditions for stopping the algorithm could be as follows:

$$(k_{iter} = k_{max}) \vee \max_{i,j} |f(x_i) - f(x_j)| < \varepsilon \vee \max_{i,j} |x_i - x_j| < \delta.$$

Note that it is necessary to investigate the convergence of the algorithm if, in step 5, the elements of population P_s will be generate as $x_k^s = \frac{1}{m} \sum_{i=1}^m x_i$, where m is a natural number. By comparing the developed algorithm with known methods, such as the genetic algorithm and the evolutionary strategy, we note that the elements of sequence P_z are intended for studying the optimum search area, and the elements of sequence P_s allow to avoid local optimums like mutations.

5   Method of Deformed Stars. Two-Dimensional Case

Let's solve the problem (2). Like a one-dimensional case, let us represent the process of solution in the form of a sequence of steps.

Step 1. Let t = 0 (iteration number).

Step 2. Generate the initial population of potential solutions $P_t = \{(x_1^1, x_2^1), (x_1^2, x_2^2), \ldots, (x_1^n, x_2^n)\}$, uniformly distributed in a rectangle $[p_1, p_2] \times [q_1, q_2]$, $|P_t| = n$.

Step 3. For all $(x_1^j, x_2^j) \in P_t$ we find $f_j = f(x_1^j, x_2^j)$, that form the population

$$F_t = \{f_j\}_{j=1}^n.$$

Step 4. Generate elements of the new population P_z. We simulate the values

$i, j = random\{1, 2, \ldots, n\}, i \neq j$. The points (x_1^i, x_2^i) and (x_1^j, x_2^j) form a segment. We will carry out the parallel transfer of this segment to the angle α and the distance a. We get new points

$$x_1^{i'} = x_1^i + a\cos\alpha, \quad x_1^{j'} = x_1^j + a\cos\alpha,$$
$$x_2^{i'} = x_2^i + a\sin\alpha, \quad x_2^{j'} = x_2^j + a\sin\alpha.$$

(5)

If at least one of points $(x_1^{i'}, x_2^{i'})$ or $(x_1^{j'}, x_2^{j'})$ lies inside the rectangle Ω (see Fig. 3), then this point is recorded in the population P_z. If at least one point is outside the rectangle, then we need to perform additional transformations that will be listed below, and then write the modified point to the population P_z. For all $(x_1^{i'}, x_2^{i'})$ and $(x_1^{j'}, x_2^{j'})$ we find $f(x_1^{i'}, x_2^{i'})$ and $f_j^z = f(x_1^{j'}, x_2^{j'})$, that form the population $F_z = \{f_k^z\}_{k=1}^n$.

Step 5. The elements of the second new population P_s are obtained in the same way as in step 4. We simulate the values $i, j = random\{1, 2, \ldots, n\}, i \neq j$. The points (x_1^i, x_2^i) and (x_1^j, x_2^j) form a segment. If $f(x_1^i, x_2^i) < f(x_1^j, x_2^j)$, then we make a turn of this segment around the point (x_1^j, x_2^j) on the angle β, in the opposite case, we make a rotation relative to the point (x_1^i, x_2^i) (see Fig. 4). In second case, the new point will have coordinates

$$x_1^{j'} = x_1^i + (x_1^j - x_1^i)\cos\beta - (x_2^j - x_2^i)\sin\beta,$$

$$x_2^{j'} = x_2^i + (x_1^j - x_1^i)\sin\beta + (x_2^j - x_2^i)\cos\beta.$$

The points (x_1^i, x_2^i) and $(x_1^{j'}, x_2^{j'})$ are written to the population P_s provided that they belong to the rectangle Ω. If at least one of the points does not belong to the rectangle Ω, then we need to make transformation from step 4. For the first case, there will be similar actions. Similarly, for all (x_1^i, x_2^i) та $(x_1^{j'}, x_2^{j'})$ we find $f(x_1^i, x_2^i)$ and $f_j^z = f(x_1^{j'}, x_2^{j'})$, that form the population $F_z = \{f_k^z\}_{k=1}^n$.

Step 6. Generation of the new population P_w is carried out in the same way as in step 5. If $f(x_1^i, x_2^i) < f(x_1^j, x_2^j)$, then we compress the segment in direction of the point (x_1^i, x_2^i) (see Fig. 5). At the same time, the point (x_1^j, x_2^j) transform to a new point $(x_1^{j'}, x_2^{j'})$, where

$$x_1^{j'} = \frac{x_1^i + x_1^j}{k}, \quad x_2^{j'} = \frac{x_2^i + x_2^j}{k},$$

k is the compression rate. The points (x_1^j, x_2^j) and (x_1^i, x_2^i) are written into the population P_w. For (x_1^j, x_2^j) and (x_1^i, x_2^i) we find $f(x_1^i, x_2^i)$ and $f_j^z = f(x_1^j, x_2^j)$, that form the population $F_w = \{f_k^w\}_{k=1}^n$.

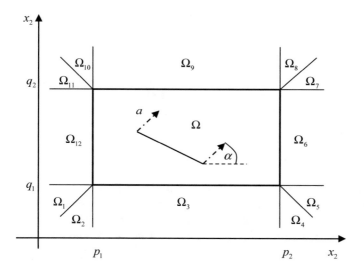

Fig. 3. Rectangle of solutions and its transformation

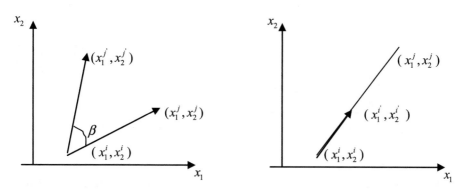

Fig. 4. A rotation   **Fig. 5.** A compression

Step 7-9. Similar to steps 6–8 of the method for one-dimensional case.

In step 4, new solutions can be out the rectangle Ω. There are 12 such cases in total. And then we need to make additional transformations so that all points of the population P_z belong to the rectangle Ω. Suppose that a point that is out Ω is $\left(x_1', y_1'\right)$. Let's consider some possible situations:

If $y_1' > q_2 \& x_1' \in (p_1, p_2)$, then $x_1' = x_1', y_1' = y_1' + (q_1 - q_2)$ (Ω_9).

If $y_1' < q_1 \& x_1' \in (p_1, p_2)$, then $x_1' = x_1', y_1' = y_1' + (q_2 - q_1)$ (Ω_3).

If $x_1' > p_2 \& y_1' \in (q_1, q_2)$, then $y_1' = y_1', x_1' = x_1' + (p_1 - p_2)$ (Ω_6).

If $x_1' < p_1 \& y_1' \in (q_1, q_2)$, then $y_1' = y_1', x_1' = x_1' + (p_2 - p_1)$ (Ω_{12}).

In other cases, the transformation is performed in the same way.

In two-dimensional case it is easy to understand why the developed method is called the method of deformed stars. We set up an algorithm for two points that are joined by a segment. In the general case, number such points may be larger and then their connection will resemble an asymmetric star, which will change the size depending on where optimal value of the function is. By approaching to optimal solution, the star will decrease, if close to the optimal value will be found for the first time, then the star will be pulled out in this direction.

6   Experiment Results

For comparative analysis of the developed method effectiveness, a genetic algorithm (GA) with elite selection of offsprings population and a tournament parent selection method was used, as well as an evolutionary strategy (ES) $(\lambda + \mu)$, where $\lambda = 10$ and $\mu = 7$ (number of offsprings for each parent) (see Fig. 6). In one-dimensional case, we investigated the Gramacy & Lee function $y = \frac{\sin(10\pi x)}{2x} + (x - 1)^4$ on the interval (0.5; 2.5).

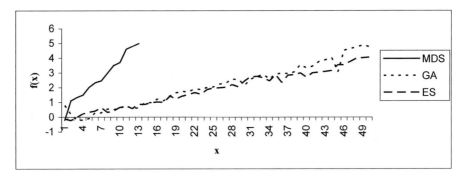

Fig. 6. Maximum value of the objective function depending on the number of iteration

For the two-dimensional case, the functions of Schwefel, Rastrigin and Grivank were used. In one-dimensional case, the problem of maximization was resolved, in the two-dimensional case, the minimization problem. Two criteria were used: the maximum number of iterations (K1) and the maximum deviation of the mean values of the target function in neighboring iterations (K2).

As shown in Fig. 6, for a given number of iterations (50), by optimizing the Gramacy & Lee function, the deformed stars method (MDS) converges to actual optimal value. The specified number of iterations is insufficient for convergence GA and ES.

The results of the MDS, GA, and ES use for optimization of the three functions are shown in Table 1. According to both criteria, the method of deformed stars was better.

Table 1. Comparative analysis of optimization methods

Functions	K1			K2		
	Deviation from the real optimum (%)			Number of iterations		
Schwefel	7	12	1,02	3007	2178	54
Grivank	8,5	9	0,8	2030	1998	70
Rastrigin	3	3,05	0,4	234	356	24

7 Conclusion

The developed method of deformed stars can be used to optimize complex functional dependencies. The obtained results showed its advantages and effectiveness, in particular, that the desired result can be obtained faster than with using other evolutionary methods. The method is parametric and it can be optimized by itself by choosing the correct values of parameters. In addition, we have considered only some variants of the method. In particular, in one-dimensional case we can consider also pairs, triple points and investigate the effectiveness of the method. In the two-dimensional case, we can do the same by moving the star towards the vertex that has the best value of target function.

An important problem is the universalization of the method and its use to optimize the functions of many variables.

References

1. Back, T., Hammel, U., Schwefel, H.-P.: Evolutionary computation: comments on the history and current state. IEEE Trans. Evol. Comput. **1**(1), 8–17 (1997)
2. Wolpert, D.H., Macready, W.G.: No free lunch theorems for optimization. IEEE Trans. Evol. Comput. **1**(1), 67–82 (1997)
3. Storn, R., Price, K.: Differential evolution – a simple and efficient heuristic for global optimization over continuous. J. Global Optim. **11**, 341–359 (1997)
4. Geen, Z.W., Kim, J.H., Loganathan, G.V.: A new heuristic optimization algorithm: harmony search. Simulation **76**(2), 60–68 (2001)
5. Yang, X.-S., Deb, S.: Cuckoo search via levy flights. In: Proceedings of World Congress on Nature and Biological Inspired Computing, India, pp. 201–214. IEEE Publications, USA (2009)
6. Cheng, M-Yu., Prayogo, D.: Symbiotic organisms search: a new metaheuristic optimization algorithm. Comput. Struct. **139**, 98–112 (2014)
7. Kaveh, A., Javadi, S.M.: An efficient hybrid swarm strategy, ray optimizer, and harmony search algorithm for optimal design of truss structure. Civ. Eng. **58**(2), 153–171 (2014)
8. Ong, Y.S., Keane, A.J.: Meta-lamarkian learning in memetic algorithms. IEEE Trans. Evol. Comput. **8**(2), 99–110 (2004)
9. Antonio, L.M., Coello, C.A.: Coevolutionary multi-objective evolutionary algorithms: a survey of the state-of-the-art. IEEE Trans. Evol. Comput. **22**(6), 851–865 (2018)

10. Cheng, R., Jin, Y., Narukawa, K., Sendhof, B.: A multiobjective evolutionary algorithm using gaussian process-based inverse modeling. IEEE Trans. Evol. Comput. **19**(6), 838–856 (2015)
11. Snytyuk, V., Suprun, O.: Evolutionary techniques for complex objects clustering. In: Proceedings of IEEE 4th International Conference Actual Problems of Unmanned Aerial Vehicles Developments (APUAVD), Kiev, Ukraine, pp. 270–273 (2017)

Detecting Flood Attacks and Abnormal System Usage with Artificial Immune System

Ivan Burmaka[1](✉)📧, Stanislav Zlobin[1]📧, Svitlana Lytvyn[2]📧,
and Valentin Nekhai[1]📧

[1] Department of Information Technologies and Software Engineering,
Chernihiv National University of Technology, Chernihiv, Ukraine
Ivan.Bourmaka@stu.cn.ua, s.zlobin75@gmail.com, Kilavv@live.com
[2] Foreign languages for specific purposes department,
Chernihiv National University of Technology, Chernihiv, Ukraine
chdtu.fld@gmail.com

Abstract. Denials of service attacks are well-known as one of the major threats in today's Internet services. It is hard to detect this type of attack, because in most cases it can not be detected with signature-based methods, because DDoS traffic often looks like normal.

This paper proposes an approach of the indirect detection of DDoS attacks and anomalies in an abuse of system resources, based on the system performance monitoring with an artificial immune system algorithm. This approach can quickly detect and warn about an abnormal server load, some types of DDoS attacks, mining scripts, botnet scripts, and ransomware.

Keywords: Distributed Denial of Service (DDoS) Attack · CPU load · Negative Selection · Abuse detection

1 Introduction

One of the most popular ways to harm servers in the global network is a DDoS attack. In this situation a fast and correct detection of this attacks is a critical task for cybersecurity specialists. So it is important to have lightweight and efficient tools for attacks detection.

1.1 Current Status of DDoS

DDoS attack sizes rise each year. If in the first half of 2016 peak attacks were only up to 579 Gbps [1], now NETSCOUT Arbor confirmed a 1.7 Tbps DDoS attack [2].

Such fantastic bandwidths are possible because hackers use amplification techniques. But such massive attacks are not often because they still need a lot

A. Palagin et al. (Eds.): MODS 2019, AISC 1019, pp. 131–143, 2020.
https://doi.org/10.1007/978-3-030-25741-5_14

of resources and in most cases just fill a server bandwidth. More often hackers try to find some vulnerabilities in a new or outdated software to make a direct influence on server resources by the less expensive attack or use a server for an attack amplification.

1.2 Distributed Denial of Service (DDoS) Attack

A Denial of Service attack is a cyber-attack in which the perpetrator seeks to make a machine or network resource unavailable to its intended users by temporarily or indefinitely disrupting services of a host connected to the Internet. A Denial of a service is typically accomplished by flooding the targeted machine or resource with superfluous requests in an attempt to overload systems and prevent some or all legitimate requests from being fulfilled [3]. If a denial of a service attack was performed from a lot of places simultaneously—this attack is called a distributed denial of a service attack.

There are two groups of DoS attacks, based on exploiting weaknesses, they bare flood and vulnerability attacks (Fig. 1) Malicious packets with some network layer protocol create a vulnerability attack. A malicious traffic is repeatedly processed by the system and after some time this vulnerable system crashes. Local Area Network Denial (LAND), Neptune, ping of death and targa3 are well unknown vulnerability attacks [5]. Flooding attacks are generated by sending a continuous malicious traffic to the servers. This malicious traffic consumes the bandwidth and makes network resources and services unavailable to the end users. The most known DDoS attacks that target the network include: UDP flood attacks, ICMP (Ping) attacks, TCP-SYN Flood attacks, TCP-PUSH-ACK attack TCP PSH+ACK Flood attacks [6].

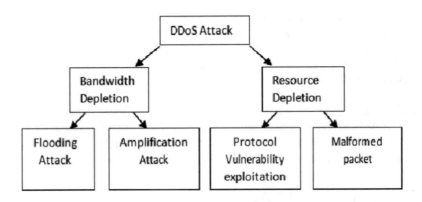

Fig. 1. Classification of DDoS attacks [4]

2   Literature Survey

This section presents the relevant literature survey on the detection of a denial of service attacks.

2.1   Signature Based Approach

This approach is also known as a rule-based or misuse-based detection. The signatures (profiles) of the previously known attacks are generated and are used as a reference to detect future attacks. For instance, a typical example of a signature would be: "there are 3 failed login attempts within 5 min" for the brute force password attack. The advantage of this type of detection is that it can accurately and efficiently detect known attacks [7], and, generally has a low false positive rate, i.e., it does not signal an alarm for a legitimate traffic.

In signature-based detection, a little variation of a known attack may affect the analysis if a detection system is not properly configured. In addition, it is unable to detect unknown or novel attacks. It is infeasible to come up with signatures for all attacks and a frequent update of a signature database is time and resource consuming. A signature-based intrusion detection can be used to detect only known external/insider attacks. And the most important disadvantage, in our case, is that this approach works very bad with DoS and DDoS attacks [8].

For example for SYN Flood attack, a hacker will spoof the source IP addresses, so the IDS will not know that this is an attack instead of just a lot of normal traffic, because DDoS requests look like a normal traffic.

But this approach is still one of the most effective versus a few other kinds of attacks, so most of Intrusion Detection Systems (IDS) use it.

2.2   Entropy Based Anomaly Detection

The analysis of random data is best done by using Entropy or Shannon-wiener index theory. It is used to measure the uncertainty or randomness present in the data. In the information theory, larger values of entropy are expected when the information variable is more random. In contrast, the entropy value is expected to be small when the amount of uncertainty in the information variable is small. If the data in the packet belong to one class, the entropy will be less and if the data in the packet belong to many classes, the entropy will be larger.

So the headers of the sampled data have been analyzed for IP and Port and their corresponding entropy is calculated. An information metric measure may be used to overcome the limitations of existing DDoS detection methods. There are three major attractions of these measures. First it helps in differentiating a legitimate traffic from an attack traffic using minimum number of attributes. The cost of this computation is low. And it can be used at various scales, in terms of number of instances taken per time window. These features are important when detecting DDoS attacks in high speed networks [9].

The threshold of a change in entropy can be defined for the detection of the DDoS attack. But this approach can cause a false positive reaction (but it is better than a false negative reaction, when the system does not detect an attack). This approach is good as a single layer of a multilayer complex intrusion detection system for DDoS detection.

3   Methodology

Most of DDoS attacks types cause an increased CPU and memory load, because a server is trying to process a huge amount of packages, so it can be an indirect sign of a DDoS attack or another abnormal work of the system, which needs attention. So we propose a methodology of the indirect detection of attacks and system resources abuse by monitoring a CPU and memory load which then processed by an artificial immune system.

This approach has a few advantages. First of all, it is a low system resource usage, because we do not check a huge amount of data, and this approach has a stable usage of system resources and does not depend on the load of the monitored system (Analysing of network packages takes much more system resources when a system is under the attack, because of huge number of packages). The second advantage is a portability, this approach can be used on almost any devices - servers, routers, workstations and embedded devices. This approach can also help to detect internal threats or system problems which makes it helpfull for detecting mining scripts, ransomware, malware used for botnets or just problems with a system setup which causes a high system resource usage.

The main disadvantage of this approach is that it can detect only problems which take affect on a CPU and memory usage. So it is not replacement of a classic IDS, but it can be good addition to a signature based IDS.

3.1   Data Capturing and Preprocessing

The most suitable for capturing a CPU usage and memory usage data is UNIX like systems, where we simply can get information about a parameter which we need from /proc filesystem or system commands which can represent the same data in more usable form, so we will use UNIX like system as an example. But it is also possible to capture the same data in windows system, but it is much more complex, because we need to use special system calls to get the performance information.

So for capturing data we are interested in we will use the following files:

- /proc/stats for CPU usage
- /proc/meminfo for information about memory
- iostat command for getting information about disk usage

For better detection accuracy we can separate a CPU load parameter to the system and user CPU load, because different types of anomalies have different fingerprints of a CPU usage. In addition to a system memory usage monitoring, we can monitor a swap partition or swapfile usage.

But having just realtime information about system resources usage is not enough for estimating a system state. We need to estimate the dynamics of parameters which we need for some periods of time, for example one, five and fifteen minutes. So we are creating the datasets which store the data for selected periods of time from the current. The next step is computing statistical parameters for these datasets - Mean value, variance, maximum and minimum values (We also can add more statistical parameters or more time periods if we need them for estimation).

After that, we will form a vector (1) from our statistical parameters, and this vector will be used for the current system state estimation.

$$\mathbf{S}(M_1, V_1, L_1, H_1, M_5, V_5.....) \tag{1}$$

Where M_n—a mean value for n minutes, V_n—a variance value for n minutes, L_n—the lowest value for n minutes, H_n—the highest value for n minutes. We need to use normalized values (2) for statistical parameters computation because our parameters have different ranges of values and it will cause problems with a distance computation (higher values have bigger impact on the distance).

$$x' = (x - MIN[X])/(MAX[X] - MIN[X]) \tag{2}$$

Where x'—a normalized value of the parameter, x—a raw value of the parameter and X—a set of raw values. Also normalized values will make it possible to have the signatures of states for some well-known problems and will make the detection more accurate.

3.2   Negative Selection Algorithm

A Negative Selection Algorithm is a core of our detecting approach and it helps to build a self-trained and adaptive system for detecting an abnormal system resources usage.

The Negative Selection Algorithm belongs to the field of Artificial Immune Systems. The Negative Selection algorithm is inspired by the self-nonself discrimination behavior observed in the mammalian acquired immune system [10]. It is an algorithm of self/non-self discrimination and mainly is used for detecting an abnormal state or non-standard pattern.

This algorithm includes two steps: generation of a detector set and detecting of new instances. At the first step we are creating a lot of different "reacting cells" - detectors which will cover the field of possible values of our state vectors. Then each detector is passing the test for matching normal state vectors. If a detector matches any of normal state vectors it is marked as a "self" detector, and it is not used for the anomaly detection (In classical "Negative selection" we need to drop self matching detectors, but in our case they will be useful in search process for marking normal state of a system). All other detectors are marked as "non-self" and matching any of this detector will raise the alarm of an anomaly state (Fig. 2). As a result we have a set of detectors. There are two approaches of detectors creating, depending on a detector size: the minimal distance to a detector center when we decided that a sample belongs to this detector. If the detectors have the same size this approach is named as Negative Selection Algorithm with Constant-Sized Detectors. If detectors are different, this approach is named as Negative Selection Algorithm with Variable-Sized Detectors [11]. Some of them represent a normal system state and some represent an abnormal state. Variable-sized detectors are more suitable in our case because they can cover the feature space more accurate and effective. It will give us a lower false rate. Now the detection system is ready to work. The Process of detection (second step) is also simple. Each sample is checked for matching to all detectors, and if we found that a sample matches the "non-self" detector - we have an anomaly (the algorithm is shown in Fig. 3).

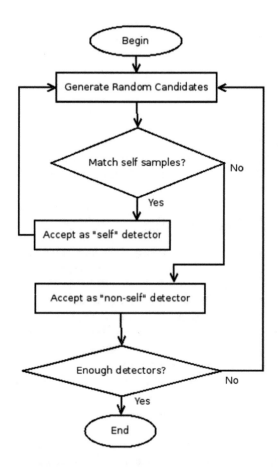

Fig. 2. Generation of detector set for negative selection

We also propose to mark "non-self" detectors with labels which can represent most popular and markable problems with a system resource abuse. It allows

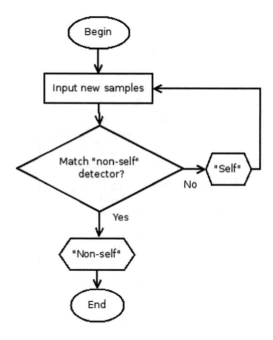

Fig. 3. Detection of new instances with negative selection

the system to tell not only about the presence of an anomaly state, but name a possible anomaly type if it has enough data for it.

Now let's move closer to our dataset. Our data is a vector of numbers, so our set of detectors will contain randomly generated and uniformly distributed vectors, in count enough to cover most possible values of real state vectors.

Instead of checking samples to direct matching with detectors we decided to search for the closest detectors with K-nearest neighbors algorithm, which is one of the simplest algorithms in machine learning for a classification task. This algorithm is a good choice in our case because it works well with multidimensional vectors and allows us to use a smaller set of detectors than we need if we want to use a direct matching. In our case will be better to use it with $K = 1$, because we want to check matching with only one detector. The algorithm of searching nearest detector and it's class ("self" or "non-self") is shown in Fig. 4.

For measuring the distance between vectors we use simple Euclidean distance (3) which works well with multidimensional vectors.

$$d(p, q) = \sqrt{\sum_{i=1}^{n}(p_i - q_i)^2} \tag{3}$$

Where n is a number of vector dimensions, p and q are vectors distance between which we are measuring. Then we choose a detector with a minimal distance to the vector as matching. If is a "non-self" detector, we have a prob-

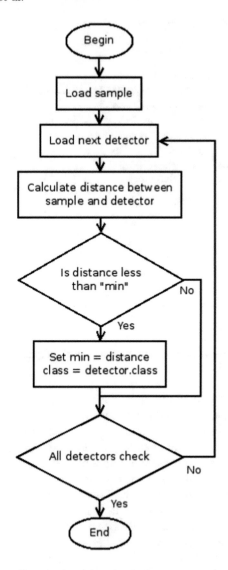

Fig. 4. Search algorithm for finding nearest detector

lem (if a detector also has a predefined problem label we even can guess which problem we have). If the detector is a "self" – then system works normally.

4   Result Analysis

This section describes experimental results and the analysis for the detection of a DDoS attack (in this case we use HTTP Flood attack—a huge amount of valid HTTP requests which create a high load of servers system) and mining script.

The main task of the experiment was to find approximate fingerprints of resource usage anomalies related to different attacks and abuse. If it is possible to find the difference between fingerprints of normal and abnormal states

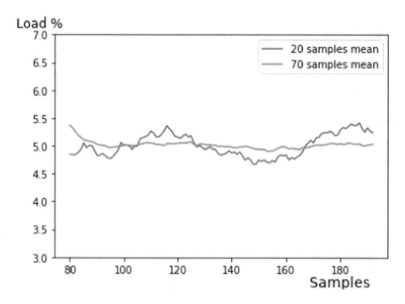

Fig. 5. CPU load on the server with low requests

then artificial immune system detectors can find these abnormal states and our approach is effective.

The normal state of the system as usually has a low CPU load and memory usage with little peaks when clients access server. Mean lines almost straight as in Fig. 5 for a CPU load and Fig. 6 for a memory usage.

Sometimes when a lot of clients access server a CPU load can get higher, in most cases, a memory usage also increases when the number of clients rise.

A mining software will cause close to 100% CPU load with few falls when the script was restarting or requesting new work (as in Fig. 7), when a memory usage is not significantly increased (most CPU mining scripts do not use a lot of memory, so the plot will look similar to Fig. 6, but the value will be higher by 1–2%) that's why it is easy to detect these types of a server resources abuse. Some brute force scripts can look the same way (so we can detect them too).

When we have a DDoS attack the CPU will be fully loaded (like shown in Fig. 7) and the memory usage will be much more higher than when our server works in a normal mode (as shown in Fig. 8), and it also will be higher than when we have a mining script running. But the exact values of a CPU and memory usage can depend on many factors, such as available resources of the target system, the software installed on a server and the resource usage in a normal state. In case of the experiment a server was a single core virtual machine, so it was not possible to load all memory by overloading a server, because of the weak CPU.

We will have the same behavior when our server is overloaded with requests, but in our case it does not matter too much. In any of these cases the system

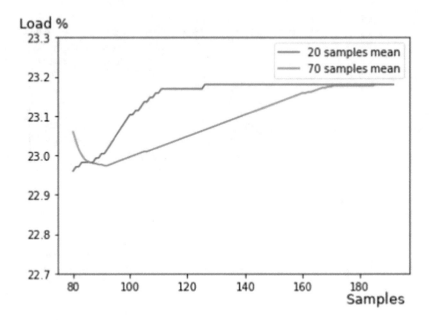

Fig. 6. Memory usage on the server with low requests

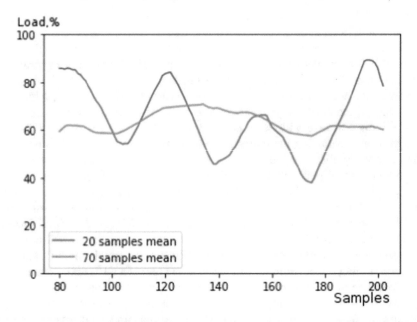

Fig. 7. CPU load on the server with mining script running

will inform us that a server can't handle requests and we need to take action to decrease a server load.

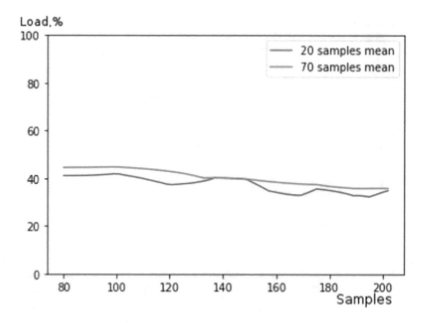

Fig. 8. Memory usage on the server underhigh load or DDoS attack

4.1   Comparison with Rule Based Detection

To evaluate the general performance of a system based on our methodology we modeled an approximate CPU load of the artificial immune system (based on the list of monitored parameters and the frequency of monitoring). Then we decided to compare modeling results with CPU load data for one of the most popular opensource intrusion detection system "Snort", which is the example of the rule-based intrusion detection system. The results of comparison are shown in Fig. 7.

As we can see in a normal state our approach uses fewer CPU resources which can help to save electricity or use this few percents of CPU time for another software. But one of the main performance advantages of our approach is that the amount of data which we need to monitor does not depend on a system load, when a rule-based system needs to analyze all network traffic, which will increase the CPU load if we have some flood attack or a lot of clients for a rule based IDS (Fig. 9).

So, detecting attacks or system resources abuse based on the artificial immune system and performance parameters evaluation is more effective, if we have a lot of network traffic or our system is not very powerful, than using rules-based intrusion detecting system.

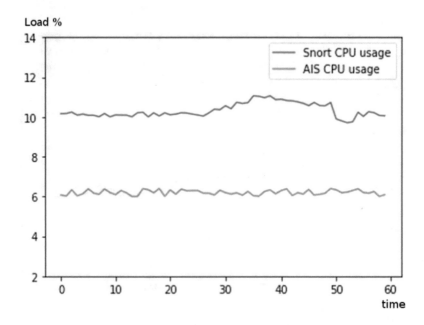

Fig. 9. Snort and artificial immune system based on IDS CPU usage

5 Conclusions and Future Work

This paper has proposed the approach and algorithm for effective detecting DDoS attacks (TCP SYN Flood, HTTP Flood) and other types of system resources abuses like mining scripts, botnet scripts, and ransomware. The proposed algorithm consists of system resources monitoring, features extraction, and detecting an anomaly with negative selection detectors. The effective detecting is based on the immune system approach where we are raising alarm on any "non-self" sample of parameters. This approach allows us to detect also unknown attacks or abuses of system resources.

There are a few directions for future work. The major task is to select a set of parameters which are optimal for detecting anomalies to minimize a vector of parameters. This will make a search of matching detectors faster. Another direction is developing an unsupervised machine learning algorithm for predicting optimal range of system resources usage for a normal system state. This will help to decrease the false positive error rate. Also it is necessary to test this algorithm on different types of DDoS attacks and abuses on different architectures.

References

1. Arbor networks releases global DDoS attack data for 1H 2016 (2016). https://www.netscout.com/news/press-release/global-ddos-attack-data
2. Morales, C.: NETSCOUT arbor confirms 1.7 Tbps DDoS attack; the terabit attack era is upon us (2018). https://asert.arbornetworks.com/netscout-arbor-confirms-1-7-tbps-ddos-attack-terabit-attack-era-upon-us/

3. NCCIC: understanding denial-of-service attacks https://www.us-cert.gov/ncas/tips/ST04-015
4. Suraparaju, V.: Taxonomy of DDoS attack (2016). https://www.researchgate.net/figure/Taxonomy-of-DDoS-attack/-4-Various-researchers/-have-given-different/-classifications/-for_fig3_309466519
5. Kshirsagar, D., Sawant, S., Rathod, A., Wathore, S.: CPU load analysis & minimization for TCP syn flood detection. Procedia Comput. Sci. **85**, 626–633 (2016)
6. Girma, A., Garuba, M., Li, J., Liu, C.: Analysis of DDoS attacks and an introduction of a hybrid statistical model to detect DDoS attacks on cloud computing environment. In: 2015 12th International Conference on Information Technology-New Generations, pp. 212–217. IEEE (2015)
7. Butun, I., Morgera, S.D., Sankar, R.: A survey of intrusion detection systems in wireless sensor networks. IEEE Commun. Surv. Tutorials **16**(1), 266–282 (2014)
8. Modi, C.N., Acha, K.: Virtualization layer security challenges and intrusion detection/prevention systems in cloud computing: a comprehensive review. J. Supercomput. **73**(3), 1192–1234 (2017)
9. Bhuyan, M.H., Bhattacharyya, D., Kalita, J.K.: An empirical evaluation of information metrics for low-rate and high-rate DDoS attack detection. Pattern Recogn. Lett. **51**, 1–7 (2015)
10. Brownlee, J.: Clever Algorithms: Nature-Inspired Programming Recipes (2015)
11. Li, D., Liu, S., Zhang, H.: Negative selection algorithm with constant detectors for anomaly detection. Appl. Soft Comput. **36**, 618–632 (2015)

Method of Encoding Structured Messages by Using State Vectors

Roman Andrushchenko[ID], Sergei Zaitsev[(✉)][ID],
Oleksandr Druzhynin[ID], and Mykhaylo Shelest[ID]

Chernihiv National University of Technology,
95 Shevchenka Str., Chernihiv 14035, Ukraine
arbamor@ukr.net, serzal979@gmail.com,
druzyninalex@gmail.com, mishell3141@gmail.com

Abstract. Structured messages are widely used in information systems for data transmission over network channels. The information transmitted goes through several stages of transformation, e.g. serialization, compression, encryption, error-correction encoding. These processes are usually considered separately from each other, following the single responsibility principle, and the corresponding algorithms usually consider the data only as a set of bits, bytes or blocks without analyzing their structure and meaning. On the one hand, this approach simplifies the process of encoding/decoding; on the other hand, it ignores a piece of information that could be used to increase the probability of success transmission. In this paper, additional method is proposed as a way of increasing the probability of successful transmission of messages through network channels by using information about the internal structure of the messages.

Keywords: Computer networks · OSI · Error-correction codes

1 Introduction

1.1 Target Settings

Communication process is an integral part of information systems. The number of communication channels quickly increases all over the world, along with their bandwidth and along with the increasing of using wireless protocols. All these things demand the corresponding level of data transfer and reception process reliability; however, modern encoding algorithms and methods usually consider outgoing data to be transmitted only as a set of bits or byte blocks without analyzing its structure [1].

On the one hand, such approach simplifies the process of encoding/decoding; but on the other hand, it ignores some piece of the information that could be used to enhance the characteristics of the data transmission process. And due to the significant spreading of wireless communication channels and rapid growth of the number of devices which have access to the Internet, more and more attention should be paid to the methods of reliable data transmission in computer networks [2].

A. Palagin et al. (Eds.): MODS 2019, AISC 1019, pp. 144–153, 2020.
https://doi.org/10.1007/978-3-030-25741-5_15

1.2   Scientific Researches and Issues Analysis

Forward error correction is used for two purposes: error detection and error correction. Objectives are achieved by adding a certain amount of redundant information (k bits) to a message of n bits length. The ratio $R = k/n$ is called the code rate. There are two classes of codes, depending on whether there is a possibility to determine clearly the rate of code: rate codes ($R = $ const), and rateless codes ($R \neq $ const). Examples of the first are BCH-codes, low-density parity-check codes, and some others [3, 4]. Examples of rateless codes are different types of fountain codes – LT codes, Raptor codes, etc. [5–7].

Several types of codes can also be combined for the sake of getting better results. Such approaches leads to new codes called "cascade codes". In details, the input data stream goes through several encoders before being sent via the channel in this case. Cascade codes are considered as one of the possible approaches of further enhancing the reliability of data transmission [8, 9].

Also, another ways of increasing data transmitting process characteristics are being researched nowadays – adaptive methods that can change the code characteristics dynamically depending on the current state of network channel along with methods that take into account: transmission channel feedback [10], features of protocol stack being used [11], network configuration [12], device types (mobile, built-in, stationary), transport (Ethernet, WiFi, Bluetooth) [13, 14].

1.3   The Goals

The purpose of this article is to explain the method of encoding information to be transmitted over a network channel, which combines the abilities of forward error codes, data compression, in combination with the advantages of statefull protocols and which takes into account the content and structure of data being transmitted. The suggested approach increases the reliability of data transmission and reduces the number of ARQs (Automatic Repeat Requests).

2   Encoding by Using State Vectors

2.1   Typical Data Transfer Preparation Process

In order to transfer some structured data between hosts of the information system, it is necessary to prepare these data firstly and convert it to a form that is acceptable for further transferring. Typically, information is stored in the device memory in a form of various in-memory structures. These structures may include tokens, lists, arrays, trees, graphs, etc. One important thing is that making a dump of such structures won't do suitable byte stream for transferring to another device, since it may contain: links to other structures (memory addresses) and unique formatted or optimized values.

In the first case, the receiver won't be able to properly decode the message because it does not have access to the sender's memory and thus it can't read the address links properly. In the second case, it is quite possible that receiver has rules of decoding certain types of data different from sender ones (for example, byte order) [15].

Based on these restrictions, serialization is a necessary step of transforming in-memory structures into a binary/text stream that can be properly interpreted by both the sender and the receiver. Other steps are optional [16].

These steps are clearly depicted below (see Fig. 1):

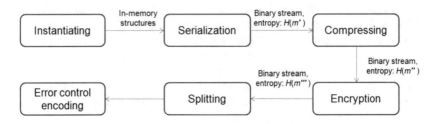

Fig. 1. Steps of preparing data to be transmitted.

Note, all these stages, except error control encoding, usually increase an information entropy H, and accordingly lead to a loss of piece of knowledge about the internal state of the system. Here, entropy is considered as Shannon's information entropy, which takes values in rage from "0" to "1" and is described by the following equation:

$$H(X) = -\sum_{i=1}^{n} P(x_i) \log_b P(x_i) \qquad (1)$$

For messages of finite length, the values x_i of the random variable X are the frequency of the occurrence of the symbol.

The last stage (error control encoding) aims inversed process: to add additional information into the message which can be used for increasing the probability of success message decoding on the receiver side. That is, ideally the process of error control encoding should know as much information as possible about the data to be encoded. However, as a result of all previous data processing processes, this information is lost. And since structured requests and messages typically contain parts that are common to a set of messages transmitted during a period of time, the hidden knowledge about structure can be considered as the current state of the system. It makes possible to use the knowledge about this state for increasing the efficiency of the data transfer process by isolating the state via creating a special statefull protocol [17].

2.2   Stateless- and Statefull Protocols

Stateless protocol is a protocol in which each participant doesn't have any information about the state of other participants. The sender transmits the data packet and does not expect acknowledgment from the recipient. To encode/decode a data fragment, enough information is contained in the data packet itself and in the encoding/decoding algorithm. Examples of stateless protocols are HTTP, UDP [18, 19].

Statefull protocol is a protocol whose members may be aware of the status of other participants. Examples of statefull protocols: TCP, Telnet. It is possible to combine statefull & stateless-protocols in an arbitrary order. An example of such combination is the stateless-protocol HTTP, which is built on top of the statefull protocol TCP.

Statefull-protocol can be transformed into stateless-protocol by turning the status vector S into input arguments $X = [x_1, x_2, \ldots, x_N]$ of data coding/decoding function:

$$f_{\text{statefull}}(t(X), S_1, S_2) = f_{\text{stateless}}(t(X + S_1 + S_2)) \tag{2}$$

Here: X – data to be transmitted, S_1 – state of the sender, S_2 – state of the receiver, $t(\ldots)$ – data transmitting function, $f_{\text{statefull}}(\ldots)$ – message decoding function of statefull protocol, $f_{\text{stateless}}(\ldots)$ – message decoding function of stateless protocol.

The amount of data transmitted over the network channel, obviously, increases in this case. The fair and inverse statement: if the state vector can be extracted from the origin message then the total amount of data transmitted theoretically can be reduced.

2.3 Extracting the Scheme of Structured Message

Structured messages have integral parts in their structure (for example: fields, key-value pairs) of some fixed or dynamic type, which often are called "tokens". Tokens' values belong to some alphabet that they can accept. Therefore, at least the following things can be analyzed and pointed out in each message:

(1) Actually used set $V = \{v_1, v_2, \ldots, v_N\}$, which is a subset of all possible values U for the current token: $V \subset U$
(2) Token identifier $i \in I$

Therefore, the message can be represented as: $m = \{i_1, \{v_{11}\ldots v_{n1}\}, \ldots, i_k, \{v_{k1}\ldots v_{kn}\}\}$, and also it can be splitted into two parts: scheme s and data d:

$$s = \{i_1, b(V_1), i_2, b(V_2), \ldots, i_n, b(V_n)\} \tag{3}$$

$$d = \{f(\{v_{11}\ldots v_{n1}\}), f(v_{12}\ldots v_{n2}), \ldots, f(v_{1k}\ldots v_{nk})\} \tag{4}$$

Here: $b(\ldots)$ – function of finding token's bounds. It isn't mandatory, and the whole set of values may be stored instead of it, but then the scheme length is increased.

$f(\ldots)$ – function of encoding the token, its purpose – extracting redundant data from the message, because it is the part of scheme s.

After splitting the message into 2 parts by the way explained above, it is possible to send the scheme s during the first connection/session, and after that – only data d has to be sent without scheme itself during all subsequent connections (Fig. 2):

The process of splitting into the schema and data must be included before compression and encryption; otherwise the complexity of finding and extracting the components of the original data packet will be significantly increased. The figure below represents this process as a "black box" with inputs and outputs (Fig. 3):

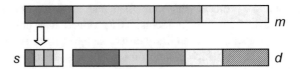

Fig. 2. Splitting the origin message into the scheme-part s and data-part d. Each color represents one token from the origin data message

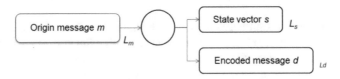

Fig. 3. Input & outputs of the splitting process

Here L_m, L_s, L_d values represent lengths of the origin message, the state vector and the encoded message respectively. These values are restricted by the following inequalities:

$$\begin{cases} L_d > L_m \\ L_m < L_d + L_s \\ H(s) > H(s+d) \end{cases} \tag{5}$$

$H(\dots)$ – information entropy. These inequalities results that the amount of data that needs to be transmitted after the synchronization of state vectors between the transmitter and the receiver is reduced by the value $\Delta L = (L_m - L_d)$.

2.4 Forming State Vector Based on Message Scheme

The result of the splitting in the form of extracted message schemes and network channel state has to be post-processed: firstly – it has to be saved in the device's memory, and secondly – it has to be transmitted to the receiver (only once for a certain period of time), thirdly – the relevant information has to be delivered to the downstream error control encoding algorithm. Comparison with existing state vectors is necessary, since the typical message scheme may be changed. Additionally, if there is used some algorithms that don't have the information about the initial data structure to distinguish the scheme, the result of their work, obviously, won't be constant in time, but it depends on the data content inside the message. The need of the third paragraph may seem unclear. Let's consider this moment in more detail. The figure below shows a modified scheme of preparing the message for further transmission:

As a result of the comparison, the state vectors should be updated if some inaccuracies were detected. State vectors contain an extracted encoded message scheme along with additional state data common to all messages for a specific time interval t. The data transmitted to the compression stage is transformed in the manner shown above in Fig. 4, and therefore, state vectors are required in order to read and interpret data correctly on the receiver side. Therefore, they are transmitted separately to the stage of forward error correction.

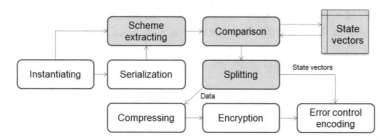

Fig. 4. Additional steps of preparing structured message for further transmitting

2.5   Message Decoding Process

Let's consider some linear (n, k)-code C. The binary stream with length of m bits in this case is divided into m/k blocks. Redundant data of length $(n - k)$ bits is added to each block. This code has a minimum distance less than or equal to the Singleton bound:

$$d = n - k + 1 \tag{6}$$

This means that the max correction ability of the code (the number of errors that it can detect and correct) is equal to:

$$t_{\max} = \frac{d - 1}{2} \tag{7}$$

The minimal distance is the distance in the n-dimensional space between the nearest codewords in the sense of the Heming metric. Let's evaluate the probability of a successful decoding of a data block with length n at a given probability of a single error p. Supposed that events lead to errors in bits are independent and random. Then the number of errors per block should be less than the critical number, which is defined as follows:

$$t \leq t_{\max} = \frac{d - 1}{2} = \frac{n - k + 1 - 1}{2} = \frac{n - k}{2} \tag{8}$$

It means that the block will be successfully decoded only in case of the number of error events is less than t. Then, according to the Bernoulli shift for repetitive tests, the probability of obtaining the number of $t \leq t_{\max}$ errors is calculated as follows:

$$
\begin{aligned}
P(t \leq t_{\max}) &= \sum_{i=0}^{t} C_n^i p^i (1 - p)^{n-i} \\
&= \sum_{i=0}^{t} \frac{n!}{i! \cdot (n - i)!} p^i (1 - p)^{n-i}
\end{aligned} \tag{9}
$$

If the information to be transmitted is considered only as a byte stream, then the Singleton bound shows the maximum correction ability of the code. However, in case of presence of additional semantic information about the structure of the message itself and information about the data contained in it, it is possible to extend the bound to some value Δd.

This approach is depicted below on Fig. 5:

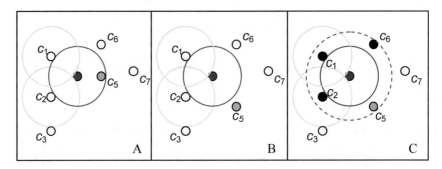

Fig. 5. Success decoding (A), failed decoding (B), success decoding if there is additional information about possible symbols (C). Solid-line circles show minimal distance between symbols. Dash-line circle – extended distance that is larger than minimal one by some value Δd.

In the ideal case, the possible amount of corrected errors increases and is:

$$t_{ext} = \frac{d + \Delta d - 1}{2} = t_{max} + \frac{\Delta d}{2} \tag{10}$$

Additional probability of successful decoding of the block increases by value:

$$P(t < \Delta d/2) = \sum_{i=0}^{\Delta t/2} C_n^i p^i (1 - p)^{n-i} \tag{11}$$

Note that this value can be considered as an additional bound for the probability of a successful decoding of the data block. The plot (see Fig. 6) shows the probabilities of absence of ARQs depending on the amount of data transmitted.

Since most of the existing encoding methods consider the input data stream as a set of bytes, the additional usage of this method, which takes into account the content of the message, leads to the extension of the boundary characteristics, thereby increasing the probability of successful decoding.

In order to select and synchronize state vectors, it is necessary to accumulate some statistics of transmitted messages. During this initial period of time, state vectors don't contain relevant information for decoding and are frequently corrected and synchronized, so the amount of data that has to be transferred increased and the protocol performance is not improved. This case is represented by the bottom line on the plot. Duration of this period depends on the intensity of the requests and the complexity of

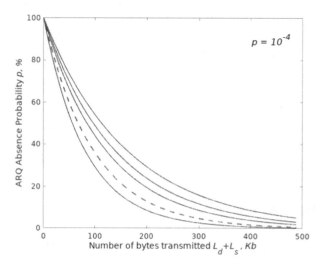

Fig. 6. Dash-line represents results on Singleton bound that can't be exceeded if message is considered as a byte stream only. Line below represents case when state vectors are not synchronized or when they can't be extracted from the set of messages due to lack of accumulated data.

the system, but in any case for structured messages it is finite. If the system accumulates enough information about the structure and patterns, then the accumulated data can be used to increase the probability of successful decoding, which is shown by the lines above the dashed one on the plot.

For the modeled system there are the following example results: without state vectors the probability of ARQs absence is 59.5% and with state vectors after full synchronization – 68.2% (54% before synchronization) for data streams with size about 50 Kb. In case of larger stream, e.g. over 100 Kb the probability of ARQs absence without state vectors is 39.3% and with state vectors – 45.4%. These values have been obtained for the following parameters: data formats – JSON, XML and MessagePack; the number of tokens per message – 20; probability of having valid data in state vector – 0.2; lengths ratio $L_d/L_m = 0.9$; single error probability – 10^{-4}, significance level $\alpha = 0.05$. Environment – Eclipse IDE, Java 9 (OpenJDK), Ubuntu 18.04, post-processing of results were performed in GNU Octave. Data messages with random content for initial tests and simulating were generated by Faker tool (https://github.com/DiUS/java-faker). Initially the message format was JSON. Converting into XML and MessagePack was performed by using Jackson tool (https://github.com/FasterXML/jackson). Along with random generated data, third-party sources were used too. They are open and presented in the table below (Table 1):

Table 1. Data sources

Description	URL
Astronauts	http://api.open-notify.org
Country demographics	http://api.population.io/1.0
Financial data of markets and GDP of countries	http://api.worldbank.org
Historical events by region	http://www.vizgr.org/historical-events
NASA database	https://data.nasa.gov
Nobel laureates	http://api.nobelprize.org/v1
Yahoo databases	https://query.yahooapis.com/v1

3   Conclusions and Suggestions

The considered method of data transmission can increase the probability of the ARQ absence due to the use of information about the message structure and its content, in contrast to standard methods that consider the input data stream only as a set of bits or blocks of bits, and also in contrast to methods that take into account the parameters of the data transmission channel, current bandwidth, etc. The proposed method is not yet independent; it can complement existing encoding/decoding algorithms. Theoretically, it can be developed up to self-use, but it is obvious that efficiency will be higher in combination with already existing and tested in practice methods.

In addition to the benefits, of course, it is necessary to consider possible disadvantages and ways of their further elimination:

(1) The complexity of the encoding/decoding algorithm is increased. This disadvantage can be partially overcome by, for example, using libraries and dynamic loading.
(2) For messages with random structure and for random data, the method will not work. That is, if it's going to transfer, for example, video stream or graphic data, the probability of finding certain dependencies in data blocks will be minimal. On the other hand, it is usually possible to admit some percentage of packet loss for media data, so there is no special need to use additional methods of forward error correction.
(3) There is a need to store state vectors in memory along with the need of state vectors synchronization. However, the amount of extra memory depends on the complexity of messages and information to be transmitted. Typically, devices that are part of embedded systems and have obviously memory limits use simple protocols and data structures to communicate, so in most cases this disadvantage also isn't critical, even for devices with limited resources.
(4) Synchronization of state vectors is mandatory part of the method that reduces the characteristics of data transmission at the beginning (during startup period). Its influence can be minimized if the algorithm of calculating the difference between the current and the modified state vector will be implemented. This will also reduce the load on the data channel.

Thus, in spite of some disadvantages, there are possibilities of their elimination and further improvement of the method.

References

1. Eady, F.: Networking and Internetworking with Microcontrollers. Newnes, Amsterdam Boston (2004)
2. Hong, S.-C., Kim, J., Park, B., Won, Y.J., Hong, J.W. Traffic growth analysis over three years in enterprise networks. In: 15th Asia-Pacific Conference on Communications 2009, pp. 896–899 (2016)
3. Kudryashov, B.D.: Fundamentals of coding theory. BHV-Petersburg, St. Petersburg (2016)
4. Novikov, R.S., Astrakhantsev, A.A.: Analysis of characteristics error-correcting codes. Syst. Inf. Process. 9(116), 164–167 (2013)
5. Kun, P., Zihuai, L., Uchoa-Filho, B.F., Vucetic, B.: Distributed network coding for wireless sensor networks based on rateless LT codes. IEEE Wirel. Commun. Lett. 1(6), 561–564 (2012)
6. Mladenov, T., Nooshabadi, S., Kim, K.: Efficient GF(256) raptor code decoding for multimedia broadcast/multicast services and consumer terminals. IEEE Trans. Consum. Electron. 58(2), 356–363 (2012)
7. Shengkai, X.: Systematic Luby transform codes with modified grey-mapped QAM constellation. Electron. Lett. 54(12), 766–768 (2018)
8. Kuznetsov, A., Serhiienko, R., Prokopovych-Tkachenko, D.: Construction of cascade codes in the frequency domain. In: 4th International Scientific-Practical Conference Problems of Infocommunications. Science and Technology (PIC S&T) 2017, pp. 131–136 (2017)
9. Shuvalov, V.P., Zakharchenko, N.V., Shvartsman, V.O.: Discrete messages transmitting. Radio Commun., Moscow (1990)
10. Azmat, A.F., Simoens, P., Van de Meerssche, W., Dhoedt, B.: Bandwidth efficient adaptive forward error correction mechanism with feedback channel. J. Commun. Netw. 16(3), 322–334 (2014)
11. Haque, A.H.M., Mondal, I.N., Ghosh, S.K., Bhotto, Z.A.: End to end adaptive forward error correction (FEC) for improving TCP performance over wireless link. In: 2006 International Conference on Electrical and Computer Engineering, pp. 569–572 (2006)
12. Yang, Z., Cai, L., Luo, Y., Pan, J.: Topology-aware modulation and error-correction coding for cooperative networks. IEEE J. Sel. Areas Commun. 30(2), 379–387 (2012)
13. Garcia, M., Aguero, R., Munoz, L., Irastorza, J.A.: Stabilizing TCP performance over bursty wireless links through the combined use of link-layer techniques. IEEE Commun. Lett. 10(3), 153–155 (2006)
14. Osunkunle, I.: AL-FEC wireless rate adaptation for Wifi multicast. In: 2018 Canadian Conference on Electrical & Computer Engineering (CCECE), pp. 1–5 (2018)
15. Tanenbaum, A.S., Todd, A.: Structured Computer Organization. Pearson, Boston (2013)
16. Bonaventure, O.: Computer Networking: Principles, Protocols and Practice. Saylor Foundation, Washington (2011)
17. Baylis, J.: Error-Correcting Codes: A Mathematical Introduction. Chapman & Hall, New York (1998)
18. Berners-Lee, T.: RFC 1945: Hypertext Transfer Protocol – HTTP/1.0. https://tools.ietf.org/html/rfc1945. Accessed 20 Apr 2019
19. Larzon, L.A.: RFC 3828: The Lightweight User Datagram Protocol (UDP-Lite). https://tools.ietf.org/html/rfc3828. Accessed 20 Apr 2019

Dynamic Assessment of the UAS Quality Indicators by Technical Diagnostics Data

Bashyns'ka Ol'ha[1], Kazymyr Volodymyr[1], Nesterenko Sergii[2(✉)], and Olga Prila[1]

[1] Chernihiv National University of Technology,
95 Shevchenka Str., Chernihiv 14035, Ukraine
bashinskaolga@gmail.com, vvkazymyr@gmail.com,
olga.prila1986@gmail.com
[2] State Scientific and Research Institute of Arms Testing and Certification
of the Ukraine Armed Forces, 1 Strilets'ka Str., Chernihiv 14033, Ukraine
cranoxy@gmail.com

Abstract. In the article a new method and software for assessment of UAS quality indicators is proposed. The definition of metrics for UAS quality indicators is given. To calculate the metrics by technical diagnostic data the method based on Bayesian belief network is used. The information structure of assessment process and main components of software application developed to support the dynamic UAS QI assessment during UAS testing are described.

Keywords: Unmanned Aerial Systems · Quality indicators · Metrics · Technical diagnosis · Bayesian Belief Networks

1 Target Settings

Unmanned aerial systems (UAS) recently have been rapidly developing [1] both conceptually and technically. The reason for this is they quite unexpectedly found as effective alternative to traditional means and technical systems in a number of areas of use.

UAS is a complicated technical system, an adequate quality assessment looks like a separate technical problem. Technical diagnostics (TD) is a significant part of the process of developing and testing UASs at the different stages of their lifecycle. Therefore, the task of rational organization the quality assessment of UAS with the help of data TD is relevant.

2 The Aims of the Article

In the pre-production period of the UAS life cycle, TD data accumulate in large volumes. At the same time, a rational evaluation of the final quality indicators (QI) through TD requires the development relevant means how to do this.

© Springer Nature Switzerland AG 2020
A. Palagin et al. (Eds.): MODS 2019, AISC 1019, pp. 154–163, 2020.
https://doi.org/10.1007/978-3-030-25741-5_16

The aims of the article are:

- to describe the action sequence for assessing the quality of UAS via the TD data to increase the validity of decisions on their compliance with the system requirements;
- to describe how BN may be used for QI assessment;
- to propose the method and means for dynamical UAS QI assessment in the process of UAS examination.

3 Actual Scientific Researches and Issues Analysis

In accordance with [2], under the test of industrial products (engineering samples), one understood an experimental determination of the quantitative or qualitative characteristics of the object tested.

In the examination process the TD data accumulate in large volumes, especially when automated systems for registration primary information are using [3]. They are not always structured enough and can be easily interpreted not always. The UAS QI definition becomes possible after processing of large data arrays only, which slows down the overall testing process. These barriers overcoming is possible by using two approaches:

- structuring of information entities to different semantic layers with ensuring of correctly organized interfaces between them;
- using the modern probabilistic mathematical methods of current TD data processing in real time. Such a mathematical apparatus is Bayesian belief networks (BN);
- using appropriate computer application.

Now more and more researchers' attention is attracted [4, 5] to the methodology of using BN to solve a variety of technical problems, especially those related to uncertainty and the need to combine expert estimates with numerical data accumulated in various databases.

BN now occupies [6] the place of one of the most productive mathematical approaches that allows flexible and adequate description of decision making by qualified experts in the diagnosis of complex systems under uncertainty [7]. Models built on these principles show themselves well in the tasks associated with incomplete and inaccurate information. With the help of BN significant advances have been made in such areas as medicine [6] (diagnosis of lymph nodes, refinement of diagnoses), automatic speech recognition systems [8], image processing, classification of data of various nature, and others.

The [4] points out that the probabilistic approach to the solution of complex technical problems based on the mathematical apparatus of BN has the following main advantages:

- the simulation results obtained by experts' knowledge and presented as the structure of the belief graph and as the form of probabilistic tables in nodes of the BN are more reliable;

- there is the possibility to adjust the models used and their parameters, taking into account the receipt of new information about the behavior of the object being studied.

Despite the fact that Bayesian networks are given a lot of attention in the world scientific literature, the principles of their construction and use are not yet sufficiently covered in domestic publications, which greatly impedes their understanding and application. In this paper the theoretical-logical and semantic substantiation the UAS QI determination by TD data is given. Also here information entities are structured and on this basis becomes possible to construct a real BN to solve the tasks formulated.

4   Metrics of Quality Indicators Assessment

Quality indicator of UAS is semantically defined as a tuple

$$QI = \langle Y, M \rangle, \tag{1}$$

where $Y = \{Y_i\}$ is a set of functions (properties) of a technical sample that are relevant to Q and which are tested during its examination;

M – the metric of the QI indicator which serves to quantify QI. In most cases, according to [9, 10], M_Q metrics are calculated as relative values:

$$M_Q = \frac{|X|}{|Y|} \text{ or } M_Q = \frac{|Z|}{|Y|}, \tag{2}$$

where $|X|$, $|Y|$, $|Z|$ means the powers (number of elements) of the sets X, Y and Z.

In the formula (2):

$X = \{Xj\}$, $X \subseteq Y$ is the set of functions (properties) of the technical sample that are performed according to the Q index during the examination;

$Z = Y \backslash X$ is the set of functions (properties) of the technical sample that are not performed according to the Q index during the examination.

Properties Y get defined during the technical samples testing through the implementation of diagnostic procedures which are components of technical diagnostics (TD).

The semantics of technical diagnostics is determined on the logical model, which, in turn, corresponds to the system of sets

$$Q = \langle T, M \rangle, \tag{3}$$

where $T = \{T_i \mid i \in (1...m), X_i \leftrightarrow T_i\}$ is a set of tests that are performed (or symptoms observed) when technical sample is examining;

$M = \{M_j \mid j \in (1...k)\}$ – is the set of QI metrics, which are calculated from the results of tests T by the formulas (2).

The logical connection between T and M can be illustrated by the incidence matrix *TM*:

$$
TM = \begin{array}{c|c|c|c|c|c|c|}
 & T_1 & T_2 & T_3 & \ldots & T_m & \\
\hline
 & 1 & 0 & 0 & \ldots & 1 & M_1 \\
\hline
 & 0 & 1 & 0 & \ldots & & M_2 \\
\hline
 & 1 & 0 & 1 & \ldots & 1 & M_3 \\
\hline
 & \ldots & \ldots & \ldots & \ldots & \ldots & \ldots \\
\hline
 & 0 & 1 & 0 & & 0 & M_k \\
\end{array}
\qquad (4)
$$

In the matrix (4) $TM_{ij} = 1$ if the metric M_j is to be calculated for the test T_i, which, in turn, can be either "Pass" or "Fail" in the simplest case.

Further, the failures $R = \{R_t |\ t \in (1...n)\}$ affect to results of tests (observations) T. The relationship between sets T and R can be explained by the RT incidence matrix:

$$
RT = \begin{array}{c|c|c|c|c|c|c|}
 & 1 & 2 & 3 & \ldots & n & \\
\hline
 & 0 & 1 & 0 & \ldots & 1 & T_1 \\
\hline
 & 1 & 1 & 0 & \ldots & & T_2 \\
\hline
 & 1 & 0 & 0 & \ldots & 1 & T_3 \\
\hline
 & \ldots & \ldots & \ldots & \ldots & \ldots & \ldots \\
\hline
 & 0 & 1 & 1 & & 0 & T_m \\
\end{array}
\qquad (5)
$$

In the matrix (5) $RT_{ij} = 1$ if the failure R_i is one of the reasons which affects to the result of test T_j.

The analysis of the TD process and the logical connections found in it allow us to construct a diagram of causes and consequences in determining QI of UAS (Fig. 2).

Thus, objects that participate in the UAS examining can be grouped logically into the diagnostic layer $\{R\}$, the effects layer $\{T\}$, and the layer of QI metrics $\{M\}$, as shown in Fig. 1. The matrices (4) and (5) act as interfaces between these layers.

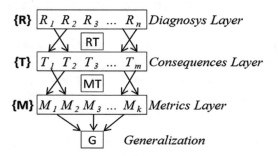

Fig. 1. Causes and consequences diagram of the TD procedure.

In Fig. 1 also there is the node of generalization of quality metrics G which is not mandatory in terms of normative documents. It looks useful technically for ensuring the possibility of final result obtaining and it can be interpreted simply in process of results analysis. Traditionally, it is calculated as a weighted sum of metrics M_i:

$$G = \sum_{i=1}^{k} \alpha_i M_i., \qquad (6)$$

where α_i is a weight of metric M_i.

The sequence of determination of UAS QI according to the TD data corresponds to the following algorithm.

1. The R, T and M sets must be determined based on the UAS technical documentation and the existing regulations. Their elements determine the course and results of the UAS assessment.
2. The matrices RT and TM are to be filled up. They determine the structure of causative relationships in the test procedure.
3. A test is conducted with two conditions:
   a. TD is performed and by this way the results of measurements and failures observation R become actualized;
   b. the results of the T test are determined and the X-sets for each metric of the M set become actualized;
4. If necessary, the generalized index G is calculated according to the customer's test method.
5. To use the model in Fig. 1 as BN, further it is necessary to define a priori probabilities for each of its objects. These a priori probabilities are determined either on the basis of the statistics of previous examinations, and/or on the basis of expert information.

The given algorithm has the following disadvantages:

1. The dimensions of RT and TM matrices can be quite large:

   $|RT| = |R| \times |T|$ and $|TM| = |T| \times |M|$. Their filling is a labor-intensive work, therefore its simplification is urgent.

2. Condition 3-b defines a slow consecutive procedure based on the RT and TM matrices obtained in point 2 above. Such an approach does not pay attention to the possibility of a logical problem decomposition, taking into account the mutual different tests independence in the UAS structure, which belongs to different subsystems of it. The method described below shows how using BN gets a solution to this problem.
3. Execution of point 5 of the above algorithm is a daunting task, since for each object it is necessary to determine a priori probabilities for the full range of common distribution of the probability of parent nodes. The situation may be much easier when:

- it is possible to determine the independence some objects from others in the model;
- when the objects of the model have a discrete distribution of values.

5 Quality Indicators Assessment with Using of BN

Let's illustrate the use of BN when UAS is being tested. For simplicity, we will consider a separate case of the UAS test, in which negative results of T3 and T4 tests became known. On the basis of this information it is necessary to estimate the most likely place of refusal.

Let's take into account the fact that, according to the UAS design, the result of the T4 test is in causal connection with the technical state of the satellite geo-stationary signals receiver (GPS), and the result of the T3 test – from the workability of both the receiver GPU and the onboard radio transceiver.

Based on these data, an example of BN, shown in Fig. 2, was created. This network is a graph whose nodes are images of probabilistic (random) variables, and the available edges reflect causal relationships between them. The absence of an edge between a pair of vertices in such a graph means the mutual independence of the probability variables corresponding to the vertex pair being considered.

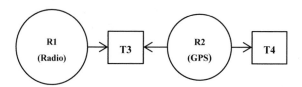

Fig. 2. Sample of the part of UAS

Let all vertices in Fig. 2 be images of Boolean random variables with values 0 or 1. The vertex R1 (Radio) on Fig. 2 depicts a random variable for which value 1 means a malfunction of the on-board transceiver, and 0 is, accordingly, its correct condition. By analogy, R2 = 1 corresponds to the fault the GPS receiver, T3 = 1 and T4 = 1 indicate the unsuccessful results of these tests. Zero values mean the proper condition of the components.

To use the graph on Fig. 2 as a Bayesian trust network, we need to set a priori probability distributions of the BN variables. These values are usually asked either by experts or by previous statistics. For our case, the distribution of a priori probabilities is given in Tables 1, 2 and 3.

Table 1. Prior probabilities for *R1* and *R2* (expert evaluation).

Value	P(R1)	P(R2)
0	0.88	0.87
1	0.12	0.13

The calculation of joint conditional probabilities (JCP) P (R1 = 1 | T3 = 1, T4 = 1) and P(R2 = 1 | T3 = 1, T4 = 1) is a sufficiently voluminous computational task, therefore let us use BayesFusion GeNIe Academic 2.3 for simulation. In this environment BN is constructed according to the topology defined by Fig. 2. The initial state of this BN is shown in Fig. 3.

Table 2. Prior probabilities *P(T4|R2)* (expert evaluation).

T4 value	R2	
	1	0
1	0.95	0.05
0	0.05	0.95

Table 3. Prior probabilities *P(T3|R1, R2)* (expert evaluation).

	R1=	1		0	
	R2=	1	0	1	0
T3= 1		0.95	0.7	0.9	0.04
0		0.05	0.3	0.1	0.96

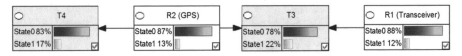

Fig. 3. Initial BN state.

Obtaining information that the *T3* and *T4* tests failed indicates that the probability of State1 in these variables became 100%. This causes an immediate change in the distribution of a posteriori JCP in all nodes of the network that have connections with them. The result is shown in Fig. 4.

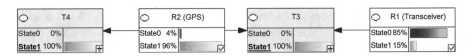

Fig. 4. The BN state after evidences *T4* = 100% and *T3* = 100%.

Let's pay attention to a sharp change in probabilities *P (R2 = 1) =* 96% and *P (R1 = 1) =* 15%. This means that in this situation, the reason for the failure in the T3 and T4 tests is the refusal of the GPS sensor, not the on-board transceiver.

The failure of one of the UAS components thus fixed thus changes the power of the sets *Y* and *X* used in (2) to determine the metrics of the quality indices.

The use of the here above method of dynamic definition of quality indicators opens the possibility of creating software applications that can improve the performance of UAS examination, shorten the time it takes to be done and the resources expenditure.

6   Process of the Dynamic Quality Indicators Assessment

Let's consider the UAS examination process the information structure of which is shown in Fig. 5.

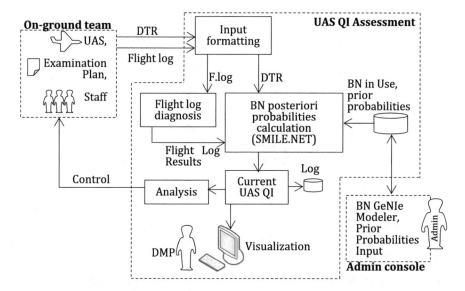

Fig. 5. The information structure of UAS QI UAS examination process: DTR – direct test's results, DMP – decision making person.

After receiving a sample for testing, and following the test plan, the testing on-ground team performs:

1. Planned sequence of UAS ground tests, in which the results of direct tests (DTR) are obtained;
2. Unmanned Aerial Vehicle (UAV, it is the part of the UAS) test flights.

During the test flights the following data are recorded:

- the results of the part of the UAS tests which performed by direct observation over the flight;
- a large number of UAS parameters that are written to the log files of its control system.

The total amount of data obtained by this way is quite large. An additional problem is that many of the named data are not suitable for direct perception by a person and need to be pre-processed and interpreted. This is especially true for log files data. Accordingly, processed and interpreted log file data yields additional test results that are very valuable, albeit indirectly.

The results of the tests obtained by the UAS testing procedure described here are often simply accumulated, and their processing is then carried out later under laboratory conditions. At the same time, the proposal of dynamical coupling in time the UAS test procedure with the current evaluation of its quality indicators looks rational. This opens up the opportunity for the test leader (the decision making person, DMP) to respond promptly to the magnitude and nature of UAS QI changes during the examination.

As a means of solving these problems, software application named TDQI was developed. The work of the application is organized by the interaction of two main modules described below.

Module for dynamic assessment of the quality indicators.

The core of this module consist of two components:

- T-LOG is a subprogram for technical diagnostics of log file data;
- Dyn-BN is a subprogram of the dynamic JCP calculation in BN, which corresponds to the complete procedure of UAS examination. Named here JCP are renewed dynamically each time after the next receipt of the results of the new test.

T-LOG Subroutine. Its input is a telemetric log file in the MAVLink format [11], which is recorded during the UAS trial flight. The subroutine extracts the required signals from the log file and generates results of the some special tests (these are some flight characteristics, parameters of GPS, radio etc.). The results of these tests are transmitted to the Dyn-BN module.

The Sub-program Dyn-BN. At the input, it forms the queue of test results which are passed to it by both the input formatting procedure and the T-LOG procedure. Then this procedure performs a sample of the next result from the queue, lists the whole set of common conditional probabilities in a given BN, and then passes these data to the procedure for converting current UAS QI values.

To calculate the JCP values in Dyn-BN, the platform-independent library of BayesFusion SMILE logical probability models (Structural Modeling, Inference, and Learning Engine) in the SMILE.NET variant is used.

UAS QI Analysis Module. It collects an array of current UAS QI values and monitors their values to be in the established limits. If necessary, a message to the on-ground team is generated about the changes in the order of the examination.

Administrative console.

This service subroutine is, at the first, for inputting and editing the BN used in the QIA evaluation module. Secondly, through it, expert estimates of a priori probabilities that are used by the QIA module are entered or edited.

Other TDQI modules perform auxiliary service functions, clear from their names.

7   Conclusions

1. Based on metrics of the UAS QI and set of TD data it is possible to construct a BN to solve the tasks for quality indicators assessment in real time during of UAS testing.
2. The proposed method and software allow quantifying the extent to which tests are expected to be used during the assessment process. This opens up the opportunity to determine the most rational and economical test methods, which include the most significant checks that give a reliable and fast result.
3. The developed software application uses the TD data in conjunction with the BN for operational control in process of UAS examination, as well as for an operational and well-founded assessment of the quality of the UAS in real time.

References

1. Herlik, E.: Unmanned Aerial Vehicles (UAVs) for commercial applications global market & technologies outlook 2011 – 2016. Technical report, Market Intel Group LLC (2010)
2. NASA-STD-8739.10. Electrical, Electronic, and Electromechanical (EEE) Parts Assurance Standard. Measurement System Identification: Metric/SI. NASA technical standard. National Aeronautics and Space Administration Approved: 2017-06-13 Washington, DC 20546-0001, p. 39 (2017)
3. Semmel, G., Davis, S., Leucht, K., et al.: NESTA: NASA engineering shuttle telemetry agent. AAAI AI Mag. 27(3), 25–35 (2006)
4. Pearl, J.: Probabilistic Reasoning in Intelligent Systems: Networks of Plausible Inference. Morgan Kaufmann, San Francisco (1988)
5. Abramson, B., Brown, J., Edwards, W., Murphy, A., Winkler, R.: Hailfinder: a Bayesian system for forecasting severe weather. Int. J. Forecast. 12(1), 57–72 (1996)
6. Onisko, A.: Probabilistic Causal Models in Medicine: Application to Diagnosis of Liver Disorders. Ph.D. Dissertation. Institute of Biocybernetics and Biomedical Engineering, Polish Academy of Science, Warsaw, March 2003
7. Akymenko, A., Bashyns'ka, O., Nesterenko, S.: Probabilistic evaluating the reliability of the control system of the unmanned aviation complex. In: DESSERT 2018 Conference Proceedings, Kyiv, Ukraine, pp. 368–371 (2018)
8. Smirnov, V.: Technology of Acceptance Control of Complex Instrumentation with Limited Resources. Ph.D. Dissertation. SU Aerospace Instrumentation, Russia, SPb (2015)
9. ISO 9000: 2007. Quality management systems. Basic Terms and Glossary. Official Edition (2007)
10. ISO/IEC 25021: 2012. Systems and Software Engineering – Systems and Software Quality Requirements and Evaluation (SQuaRE) – Quality Measure Elements (2012)
11. MAVLink Developer Guide Homepage. https://mavlink.io/en/. Accessed 25 Apr 2019

Multi-agent Monitoring Information Systems

Svitlana Kunytska$^{(\boxtimes)}$ ⓘ and Serhii Holub ⓘ

Cherkasy State Technological University, Cherkasy, Ukraine
kunitskaya33@gmail.com, s.holub@chdtu.edu.ua

Abstract. The paper proposes to implement the methodology for the creation of multi-level intelligent monitoring information systems by employing an agent approach with the use of cloud technologies. Information technology of the multilevel intelligent monitoring is implemented in the form of a multi-agent monitoring information system, which is built for the mobile provision of knowledge for each individual decision-making process. Specially created intellectual agents carry out observation and formalization of the emergencies in heterogeneous objects at appropriate levels of monitoring. Functional monitoring aggregation processes carried out in accordance with the global debt arranged by the customer, the genetics of the intelligence agencies of the required classes are studied multiple times for modeling agents and coordination of their interaction, implemented at the upper, server-level management of the multi-agent monitoring information system. The client level contains a multitude of agents and communication tools with them. Using intelligent agents for solving local information retrieval tasks can increase the diversity of the multi-level intelligent monitoring process. It proposed to apply the process of the aggregation of the multi-agent monitoring information system functions at each level of the monitoring and to implement typical units in the form of intellectual agents guided by multi-layered models. Interaction of the intellectual agents has been formed situationally as typical processes of horizontal formation and vertical connections of the method of the ascending synthesis elements of the multilevel structure of the multi-agent monitoring information system. Agents-classifiers, agent-identifiers, predictor agents are identified and described among the agents for data conversion. It is proposed to adopt the method of the synthesis multilayered models to the peculiarities of their use in the structure of the intelligence agents. It was given an example of using one of the methods of the multilayered modeling in the structure of the agent-identifier of the functional dependencies.

Keywords: Multi-level monitoring · Information technology ·
Intelligent agents · Multi-layered modeling

1 Introduction

Intelligent monitoring is an information technology providing knowledge of the decision-making processes through the organization continuous observations, processing, and transformation of their results. Monitoring is ordered for providing knowledge of the implications of managing the effects for each decision-making process individually.

© Springer Nature Switzerland AG 2020
A. Palagin et al. (Eds.): MODS 2019, AISC 1019, pp. 164–171, 2020.
https://doi.org/10.1007/978-3-030-25741-5_17

Using agent approaches in creating technologies for multi-level monitoring of the complex objects allows us to obtain a new methodology for constructing monitoring intelligent systems. The use as such systems is appropriate in cases when there is a need to consolidate knowledge about the properties of heterogeneous monitoring objects in order to overhaul the characteristics of the development of the situation as a result of the use of managerial influences by the decision maker.

The presence of a large number of intelligent agents, created for observation tasks, self-study, classification, forecasting, allows them to use in monitoring technology after appropriate refinement and their adaptation to new requirements. The multileveled of the monitoring technologies is due to the complexity of tasks in predicting the consequences of applying control effects when making decisions in a given subject area. Unlike existing multi-level monitoring systems, multi-agent monitoring systems use hierarchical combinations of the interactive agents that solve local informational transformation problems and based on multilayered methods for synthesizing monitoring object model.

2   Overview of the Methods and Tools for Multi-agent Monitoring

Information technology of the multilevel intelligent monitoring consists of three stages [1]: 1. Formation of the monitoring structure. 2. Testing of the information conversion process. 3. Use of the monitoring intellectual system. At each stage, a separate action situation is performed. These actions are subject to a bath-set.

The central procedure is to construct a model of the monitoring object based on observations for him. On the picture provides a functional diagram of the process of synthesis of models [1], which implements the multiple algorithm GMDH [2] (see Fig. 1).

An important factor is the list of controlled signs [3]. Informative results of the observations should be sufficient to construct a useful model by existing means. Thus, the model synthesizer is a separate unit that can be implemented in the form of an agent.

Models are used as information conversion algorithms. They allow us to transform an array of the numerical characteristics monitoring objects into an array of the indicators their states [4]. It means it is another unit with some functions that can be implemented in the form of an agent.

In [5] described information technology synthesis of the multiple models that can be applied in the structure of the intellectual agent. This technology, used for any information numerical mass, reveals a drawback for the synthesis of disposable, adequate models. To overcome this "limitation of one-year phenomena" [2], one model of the simulation - ring methods and obtained models form a multiple-current local functional - multiple-model.

The presence of a large number of the intelligent agents, created for observation tasks, self-study, classification, forecasting, allows them to be used in the monitoring technology after appropriate refinement and their adaptation for new requirements. The results of the information search presented in [6], in [7] and other similar publications,

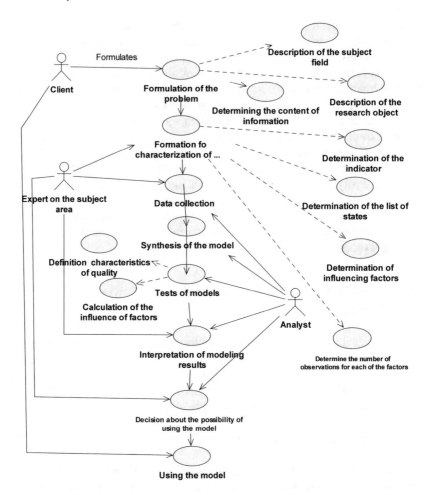

Fig. 1. Algorithm for the functioning the synthesizer models.

allow highlighting several areas for the development of intelligent agents that can be used in the monitoring intelligent systems. In particular, for each function performed in the intelligence system, a separate type of agent is developed that can adapt to the change in the external environment.

In [8], distinguish interface agents, administration, analysis and calculation of data. At the same time, the author [6] highlights the typical structure of the multi-agent system, which is based on typical agents.

A non-typical is a model that defines the functionality and efficiency of the agent. The effectiveness of the tasks performed by the agent depends on the model containing the knowledge. Using traditional knowledge bases, such as semantic networks, reduces the mobility and adaptability of the agent. Therefore, research aimed at increasing the mobility and adaptability of the intellectual agent's structure to changes in the environmental conditions is relevant.

3   Agency Approach to the Construction of the Monitoring Intelligent Systems

The use of the agent approaches in the creation of multilevel technology for complex objects complexity allows us to obtain a new methodology of the constructing monitoring intelligent systems. It is proposed, after the formation of a functional monitoring scheme, to aggregate the structure and implement the units in the form of intellectual programs. The classes of agents, which implement the monitoring process, and the list of the elements in each class are determined by the agreement of the results with the customer on the content of the information to be obtained by monitoring results. The analysis of the functional scheme the monitoring process allows to identify several typical classes of the agents and determine the typical methods for coordinating their interactions. In Fig. 2 provided information technology (IT) multi-agent monitoring, built using an agent-based approach (see Fig. 2).

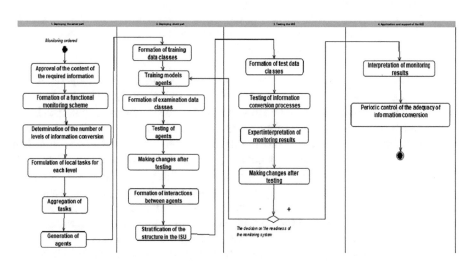

Fig. 2. Information technology of multi-agent monitoring.

Intelligent monitoring system implemented on the client-server technology. In this case, the server part combines agents that implement the functions of the synthesis models, generation of other agents and coordination their interactions. The client part contains tools for engaging with intelligent agents, in particular for generating an input array for learning models that control the work of the agent. Information technology contains 4 stages. At the first stage, the server part is deployed and typical procedures for constructing a functional monitoring scheme are implemented in accordance with the content of the information agreed with the customer.

The content of the information to be obtained by monitoring results is determined by the customer depending on the subject area. The use of such systems is appropriate in cases where there is a need to consolidate knowledge about the ownership of the heterogeneous monitoring objects in order to miscalculate the characteristics of the development the situation because of the use control effects by the decision maker. To do this, we construct a global functional dependence.

For ensure mobility, it is proposed to implement the server part of the system in a cloud, combining the agent approach to the implementation of the multi-level monitoring technology with cloud services. It is proposed to implement the model synthesizer in the form of a cloud service, which will be available to agents in case of need to resynthesize their models. File storage is used to store input data arrays for model training.

With cloud resources, the monitoring system can be deployed expeditiously for any customer, taking into account the global function of the system. The Model Synthesizer is built around the cloud-based PaaS (Platform as a Service) resource-platform as a service. There is no need to build specialized software complexes for a new monitoring system. Thanks to the cloud-based interaction model, which is classified as SaaS (Software as a Service) - the software as a service, once developed software modules can be used promptly to construct monitoring intellectual agents according to a new assignment that has determined customer.

The client part of the monitoring intelligent systems is formed at the second stage of the IT. The central process at this stage is to train models of the agents that have already been generated and belong to native classes and are elements of the same class for their ability to solve the already defined tasks of the inform transformation. For example, the class of agents "Information Converters" contains elements: cluster agent, agent-classifier, agent-predictor, agent-identifier of functional functions. These agents are intended to solve typical tasks of the transformation of the information contained in the array of numerical characteristics of the monitoring results: clustering of the observation points, classification of the state of monitoring objects, prediction of values of numerical characteristics the state indicators, identification of functional dependencies of the indicators states from the characteristics of the influences different nature and other typical problems, which are defined at the first stage in the formalization of local monitoring tasks. Determining the sequence of use these agents in data processing allows us to transform information from the formulas of the array of numerical characteristics to the form of analytical dependencies or other classes of models, such as neural networks.

Multi-layer technology allows us to obtain useful models in the conditions of insufficient in formativeness of the input data array. These models are constructed in the form of a function that contains in its structure the functions the model of the lower levels obtained by the completed algorithms.

The most common method for synthesizing multilayered models is recirculation [9] when the simulated metrics are received as an additional variable in the input array. Another well-known method of this class is the model calibration [1] - the model obtained by the completed algorithm is included in the structure of the model of the next layer for example regression of the form:

$$z = b_0 + b_1 y + b_2^2 y + b_3^3 y \tag{1}$$

where z - functional (multilayer model), y - model of the lower layer, obtained by the completed algorithm.

In Fig. 3 provides a functional diagram of the multilayer model synthesis method, called "Multimodeling".

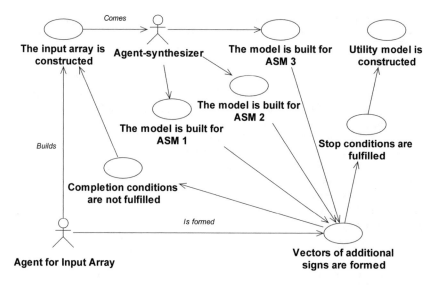

Fig. 3. Synthesis of multilayer models by the method of the multimodeling.

Each of the layers of the functional multi-model is formed by synthesizing the models of the same indicator by several methods in parallel. Modeling results are added to the input data array as additional variables. The process ends with a minimum gradient value criterion of the quality of models.

For the test the method of multilayer synthesis agent models belonging to the class "information converter" and solves the problem of identifying functional dependence, a model experiment was conducted. The dependence of the blood diseases of the population of the Cherkasy region on the concentration of harmful substances in the air of the residential zone during 2016–2018 was determined. Table 1 lists the modeling variables.

For the synthesis of agent models used:

(1) Multi-row algorithm GMDH [2].
(2) Stepanenko's algorithm [10];
(3) The method of the different elements [1].

The results of model testing are presented in Table 2.

The adequacy of the models were estimated by the value of the mean square deviation. According to Table 2, the adequacy of the model increases. The mean square deviation decreased by 16.83%. As the number of the layers increases, the speed of model synthesis increases.

Table 1. Numerical characteristics of illnesses of the population.

Characteristic	Comment
Time of observation	2016–2018
The place of observation	Cherkasy and regions
(3) nitrogen oxides (NO); (4) sulfur dioxide (SO2); (5) carbon monoxide (CO); (6) hydrocarbons (Cn (H2O) n; (7) manganese dioxide (MnO2); (8) ammonia (NH3); (9) soot (SiO2); (10) toluene (C6H5CH3); (11) butyl acetate (CH3 (CH2) 3OOCH); (12) acetone (CH3COCH3); (13) light organic compounds (VOCs); (14) calcium oxide (CaO); (15) xylene (C8H10); (16) ethyl acetate (CH3COOC2H5); (17) tetraethyl lead ((C2H5) 4Pb); (18) gasoline (C5H12); (19) phenol (C6H5OH)	The concentration of substances in the air of the living zone

Table 2. Test results of multi-layered multi-models.

Model	Average quadratic deviation of the simulation results from actual values disease, diseases/10,000 people	Time synthesize layer patterns, c
1	1762,41	37
2	1531,41	17
3	1466,01	11
4	1465,63	7

4 Conclusions

The combination of the agent approach and cloud technologies in the formation the information systems of multi-level monitoring allows to obtain a new methodology of multi-agent monitoring. It has been experimentally proved that the use of multilayer methods of synthesis agent models allows improving the adequacy of the process of informational transformation.

References

1. Golub, S.: Multi-level simulation in environmental monitoring technologies. View. From. ChNU named after Bogdan Khmelnitsky, Cherkasy, pp. 220 (2007)
2. Ivakhnenko, A., Stepachko, V.: Impulse immunity modeling. Scientific Opinion, Kyiv, pp. 216 (1985)
3. Golub, S.: Application of technology "Focusing emphasis" in information modeling in monitoring systems, Kyiv, vol. 4, pp. 45–47 (2006)
4. Golub, S.: The technology of information modeling taking into account the state of the object. Cherkasy State Technologica. Univ. **1–2**, 46–47 (2007)

5. Golub, S., Burleyy, I.: Multilayer data conversion in information systems of multi-level monitoring of fire safety. Collection of scientific works of Kharkiv University of Air Forces. Kharkiv University of Wind Power Names named after Ivan Kozhedub, Kharkiv, pp. 246–251 (2014)
6. Yaremenko, V.: An overview of multi-agent systems for data mining intelligence. Series Tech. Sci. 3(2), 47–55 (2018)
7. Leskin, V., Saprunova, K.: Intelligent agents in constructive environment-high. Electron. Commun. 2–3, 255–259 (2009)
8. Serrano, E., Rovatsos, V., Botia, J.: Data mining agent conversations: a qualitative approach to multiagent systems analysis. Inf. Sci. 230, 132–146 (2013)
9. Golub, S.: Using analogy in the design of multi-level information modeling technologies. Bull. Cherkasy State Tech. Univ. 3–4, 69–71 (2007)
10. Golub, S.: Simulation of objects of environmental monitoring by Stepanenko. Electron. Control Syst. 4(10), 165–168 (2006)

Traffic Lane Congestion Ratio Evaluation by Video Data

Inna V. Stetsenko$^{(\boxtimes)}$ (ID) and Oleksandr Stelmakh (ID)

Igor Sikorsky Kyiv Polytechnic Institute,
37 Prospect Peremogy, Kiev 03056, Ukraine
stiv.inna@gmail.com, stelmahwork@gmail.com

Abstract. The qualitative collection of traffic information is an important task. It allows to make adequate decisions at the design stage, reconstruction, repair of roads, and also helps to manage traffic. The quality of traffic control systems depends on the accuracy of determining parameters of traffic flow. The main parameter is intensity. In this article proposed to consider the problem of determining the intensity not as a problem of the classification of objects, but as a problem of segmentation. A software implementation of a method for determining the intensity of traffic, which based on a neural network and analysis of information from video cameras of traffic monitoring, has been developed. The neural network was trained on data that was obtained using the method of motion detection. It is proposed to use the TLCR (Traffic lane congestion ratio) parameter instead of AADT (Average Annual Daily Traffic) for more accurate transmission of information about the current traffic situation.

Keywords: TLCR · AADT · Traffic intensity · Neural networks · OpenCV

1 Introduction

Every year there are more and more cars on roads that leads to an increase of traffic intensity. One of the reasons for traffic jams in cities is that the roads have not been designed for the ever-increasing number of vehicles. In addition, widening of the street, especially in places of the crossroads, demands enormous financial expenses from a city budget. A huge shortage of the city's road network leads to traffic jams if an accident or road repair happens, and a driver doesn't risk getting stuck only at night. Because of limited space for building new infrastructure including limited quantity of resources the only way to solve the problem to increase quality of traffic is to improve a system of management. The system should manage the parameters of traffic flows as intensity of traffic or other indicator that shows current quantity of transport on the roads or another conditions (car crash, repair work).

Currently, data of average annual daily traffic (AADT) are used for optimization of the traffic control system's parameters. Average annual daily traffic is a total volume of vehicle traffic on a highway or a road per year divided by 365 days. AADT is a useful and simple measurement of how busy a road can be. AADT is one of the generally accepted indicators of traffic used by road planners and engineers. The article [1] shows the situation that did not affect the value of AADT, but affected on the number of traffic

© Springer Nature Switzerland AG 2020
A. Palagin et al. (Eds.): MODS 2019, AISC 1019, pp. 172–181, 2020.
https://doi.org/10.1007/978-3-030-25741-5_18

jams, which means the necessity of involving the new parameter that will determine the volume of road traffic. Besides, AADT has such weaknesses:

- Incorrect determination in the case of traffic jam (the number of passing cars will be the same as in the case of no traffic jam).
- Vehicle size is not considered.

In this article, instead of well-known AADT, the TLCR parameter computed using information obtained from traffic cameras is proposed. This parameter is intended to estimate road congestion of one lane. We define it as the mean of the values of pixels placed in the image area that is corresponded to one lane. The properly determined and predicted value of TLCR parameter will help the control system to find optimized parameters and to avoid the traffic jams. Optimization can be performed by applying the evolutionary algorithm and traffic simulation model. The image processing technology for the estimation of TLCR parameter is developed.

Section 1 introduce to the problem of measurement of traffic volume. The second section describes related works. The next section represents the methods for video detection of vehicles. Section 4 describes presents experimental results. The last section summarizes the main results of the research.

2 Related Works

In work [2] the usage of the noise from cars as a key parameter by which traffic intensity can be determined is proposed. The algorithm for calculating intensity from the acoustic radiation of a car based on fixing the amplitude of the acoustic signal consists of the following steps:

- Record the acoustic characteristics of the traffic flow.
- Convert an audio file to an array.
- Filtering.
- Construction of the envelope.
- Processing using the moving average algorithm.

To the number of cars driving through is incremented by one if the value of derivative has had a positive value, which is more than the threshold, and then a negative value, which is lower than the threshold. It is possible to change the threshold value of noise, which determines the passing car.

The authors of work [3] suggest using short-term observations to determine the traffic intensity. A formula for recalculating the hourly intensity of road traffic into average annual is proposed. The coefficients of the transition from the hourly traffic to the average annual are determined. The periods of observation of traffic intensity are justified.

In the article [4] a neural network is used to identify such objects as bicycles, cars, road signs. Data from video cameras are used. The implementation consists of three different detectors that determine the presence of objects of the corresponding classes in the images. For object classes with a large intra-class variation like cars, which appearances and shapes change significantly as viewpoints change. In order to deal with these variations that paper presents an object subcategorization method which aims to cluster the object class into visually homogeneous subcategories. The subcategorization method applies an unsupervised clustering to one specific feature space of the training samples to generate multiple subcategories. This method simplifies the original learning problem by dividing it into a few subproblems and improves generalization performance. KITTI dataset is used to train a neural network to detect cars. This dataset is a recently proposed challenging dataset which consists of 7481 training images and 7518 test images, comprising more than 80 thousands of annotated objects in traffic scenes. Fast speed is one of the advantages of this approach. In this research, we consider the implementation of the neural network to determine traffic intensity. The significant difference from previous research is the use of a neural network that implements task segmentation. The neural network is trained using data generated by motion detection methods.

3 Methods

In order to determine TLCR it is necessary to detect transport. There are a lot of methods to detect vehicles crossing the place of the road. We can conditionally divide them into two groups: invasive and noninvasive. Invasive methods propose intervention of road surface using technical devices. Noninvasive methods allow us to use the equipment and to use software to process incoming information. Listed solutions (as invasive as noninvasive) use specialized hardware which has a number of restrictions such as high price, difficult supporting, and introduction to short life term. Further, in this research, the systems of video detection are considered. They are prospective enough from the point of practical using and for the reason that only this way we can determine TLCR.

While working with cameras we face the following problems:

- The camera saves the video with a one-minute interval. In this case, it has a delay and therefore there is a one-second gap between the first shot of the following video and the last shot of the previous one.
- The problems of determining cars in road lanes.
    - Depending on camera the high vehicles can be determined on the lane they are not moving.
- Cars allocation integrity problems
    - Windscreens.
    - Shadow.
    - Car color.
    - The light of front lights at night.

There are two main approaches to build a system of video detection of vehicles. The first one is to search and track the traffic area with following identification of these areas as vehicles [5–7]. Implementing the first approach the operation of the system can be divided into such stages:

- Allocation of traffic area on the current shot of the video.
- Tracking traffic area on a few following shots.
- Building traffic trajectory for every area is assured. It should be admitted that as a result of overlaps every area can include the groups of objects.
- Dividing objects which belong to one traffic area. Classification of vehicles. Determining the length and the width of the object and other parameters that can be determined based on the received consistency of traffic area and information of road camera.
- Tracking vehicles. On this stage, building trajectory of vehicles movement is assured from identity moment to their exit from camera's side.

Currently, most of the systems operate according to the given scheme [8]. It should be admitted that the scheme based on search and tracking traffic area operates only in case of a fixed camera and permanent background. In natural conditions the camera is always under the influence of wind or downfall, that is why immobility is unreachable. As a result, program video stabilization is implemented. Typical scheme of video detecting systems via locating and tracking objects is not dependent on the types of detected objects and consists of the following procedure:

- Extraction of a video shot from image data
- Adaptation to the received shot
- Searching the object location in the image, evaluation of the authenticity of the location in the given area.
- Tracking the found objects
- Analysis of the searching and tracking results.

3.1   Search Method of the Objects in the Image

Two methods of object detection in the image are studied and tested out: two shot comparison method and bioinspired method.

Two Shot Comparison Method. In the plainest variant of this method search for differences results in examining every pixel in the first image and checking if the pixel is visible in the second image. The problem of this approach is that it only designates the change without measuring it. There is no difference whether the pixel became a bit darker or it has completely different color. To solve this problem filtration method "Gauss Blurring" has used.

Movement Detector Based on Bioinspired Module OpenCV. OpenCV library contains Retina-class had a spatiotemporal filter of two information channels (parvocellular pathway and magnocellular pathway) of the retina model. We are mainly interested in magnocellular channel which is already a motion detector itself. The only thing we need is to receive coordinates of an image area where the motion is detected and somehow react on interference which appears when movement detector shows a static image. The implementation consists of the following stages:

- Transformation of the input image from RGB into "greyscale".
- Start of the "magnocellular" module.

- Examination of the medial filter from entropy in order not to react on noise when there is no motion. Value 0,60 was used in the research.
- Comparison of the result with threshold requirement 90.

Methods described above were examined on practice to identify traffic intensity [9] (Table 1).

Table 1. The results of traffic intensity identification.

	Number of vehicles	Received number of vehicles	Time	Intensity	Intensity nominal error	Relative error
Comparison of shots	468	547	3	156	182.33	16.88%
Bioinspired module	468	490	3	156	163.33	4.70%

Thus, disadvantages of the methods based on movement detection:

- Motionless transport is not considered but at the same time, it influences traffic situation by making drivers of other vehicles drive to the adjoining lane to take a detour.
- Impossibility to detect transport in real time because of slow operation speed.

3.2   Neural Network

To overcome described above disadvantages it was decided to use neural network and to use the method that showed itself in the best way (method of movement detection based on bioinspired module OpenCV) to form data cluster to instruct this network. The model of epy chosen neural network is the suppressed version of deep education architecture called U-net. U-net is network architecture coder-decoder type for image segmentation. The article [10] represents the approach for medical images segmentation. However, the presented approach can be used for other segmentation tasks including ours. U-net network ability to operate with a small amount of data and without special demands to the size of the input image makes it a strong candidate for image segmentation tasks. The scheme of U-net model we can see in Fig. 1.

Two optimization algorithms were used to train the neural network: Adam and SGD. Adaptive moment estimation combines the idea of the accumulation of motion and the idea of a weaker updating of the scales for typical signs. Adam scales the gradient value. In order to know how often the gradient changes, the authors of the algorithm suggest using the average non-centered variance. The main advantage of Adam is that it allows you to achieve good results very quickly. The Adam method converts a gradient as follows [11]:

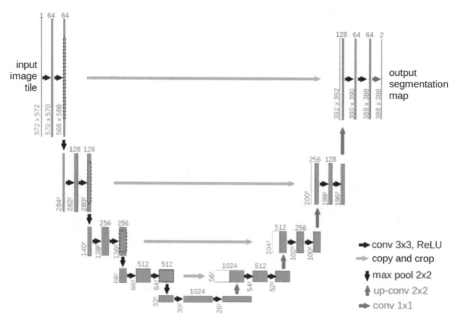

Fig. 1. U-net model.

$$S_t = \alpha \cdot S_{t-1} + (1 - \alpha) \cdot \nabla E_t^2, S_0 = 0,$$
$$D_t = \beta \cdot S_{t-1} + (1 - \beta)\nabla W_{t-1}^2, D_0 = 0,$$
$$g_t = \frac{\sqrt{D_t}}{\sqrt{S_t}} \cdot \nabla E_t, \tag{1}$$
$$\nabla W_t = \eta \cdot (g_t + \rho \cdot W_{t-1}) + \mu \cdot \nabla W_{t-1}.$$

where η is the learning rate, ∇E_t is loss function gradient, μ is moment ratio, ΔW_{t-1} is changing weights in the previous iteration, ρ is regularization coefficient, W_{t-1} is weights on the previous iteration, $\alpha = 0.999$, $\beta = 0.9$, S_t is estimates of the first moment (the mean), D_t is estimates of the second moment, g_t is gradient. Training on a data set consisting of 10000 images was made with the following parameters: the number of epochs is set to 5, the validation split is set to 0.1, the metric value is set to 'intersection over union'.

In order to train the network, it has been decided to use the method SGD (Stochastic Gradient Descent). SGD is one of the optimization methods, namely the first-order optimizer, which means that it is based on an analysis of the gradient of an object. Therefore, from the point of view of neural networks, it is often used together with Backpropagation for effective updating. To calculate SGD, you need to calculate the gradient of the model and here Backpropagation is an effective method for calculating the gradient. On the neural network diagram, you can see the number of layers. In order to improve the network, it was decided to cut the input images into segments of 128 by 128 pixels (Fig. 2). In Fig. 3 you can see the scheme of filtering which occurs on each layer of the neural network [12].

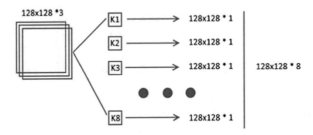

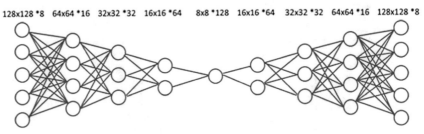

Fig. 2. The structure of neural network.

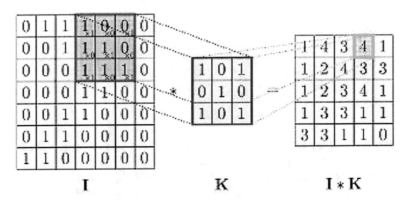

Fig. 3. The scheme of filtering.

Traffic lane congestion ratio (*TLCR*) is defined as

$$TLCR = \frac{\sum_{i \in A} V_i}{|A|} \qquad (2)$$

where i is an index of the pixel which is an element of subset A containing only indexes of pixels that describes the traffic line, $|A|$ is the cardinality of the set A, V_i is the value of the pixel, $V_i \in \{0,1\}$.

4 Experimental Results

In order to show the efficiency of transport detection using a neural network, it was decided to conduct an experiment in which we compared the bioinspired method and neural network method. Figure 4 shows an image from the surveillance camera located in the city of Kyiv. In Fig. 5 you can see the result of the neural network.

Fig. 4. The original image from camera.

Table 2 shows a comparison of the results of the use of traffic detection methods for the problem of determining the intensity of traffic flow.

Fig. 5. The result of work of neural network.

Table 2. The results of traffic intensity identification.

	Number of vehicles	Received number of vehicles	Time	Intensity	Intensity nominal error	Relative error
Bioinspired module	817	873	3	272	291	6.85%
Neural network	817	851	3	272	283,66	4.16%

5 Conclusions

A comparative analysis of methods for determining the intensity of traffic flow is made. A software implementation of the U-net neural network is developed. It is proved experimentally the advantage of using the neural network for the problem of determining the traffic intensity or TLCR. High-quality processing of information getting from surveillance cameras located at crossroads is a powerful source of information for traffic control systems to prevent adverse situations such as stopping traffic due to an accident or repair of the road. In the future, it is planned to conduct experiments and a comparative analysis of the effectiveness of the use of TLCR parameters and traffic intensity.

References

1. Johnson, G., Johnson, A.: Bike lanes don't cause traffic jams if you're smart about where you build them. FiveThirtyEight (2014). https://fivethirtyeight.com/features/bike-lanes-dont-cause-traffic-jams-if-youre-smart-about-where-you-build-them. Accessed 25 Apr 2019
2. Frantsev, S.M.: Algorithm for calculating the intensity of the transport stream based on fixing the amplitude value of the acoustic emission of a car. Eng. J. Don **2**, 1–6 (2017)
3. Malahov, R.S., Aleksikov, S.V.: Estimation of traffic volume by short-time observation. Bull. Volgograd State Univ. Archit. Civ. Eng. Ser. Constr. Archit. **49**(68), 92–98 (2017)
4. Hu, Q., Paisitkriangkrai, S., Shen, C., Hengel, A., Porikli, F.: Fast detection of multiple objects in traffic scenes with a common detection framework. IEEE Trans. Intell. Transp. Syst. **17**(4), 1002–1014 (2016)
5. Kim, Z.W., Malik, J.: Fast vehicle detection with probabilistic feature grouping and its application to vehicle tracking. In: Proceedings Ninth IEEE International Conference on Computer Vision (ICCV 2003), vol. 1, pp. 524–531. IEEE, Nice (2003)
6. Sivaraman, S., Trivedi, M.M.: A general active learning framework for on-road vehicle recognition and tracking. IEEE Trans. Intell. Transp. Syst. **11**(2), 267–276 (2010)
7. Tsai, Y.M., Tsai, C.C., Huang, K.Y., Chen, L.G.: An intelligent vision-based vehicle detection and tracking system for automotive applications. In: Proceedings of the IEEE International Conference on Consumer Electronics, vol. 1, pp. 113–114 (2011)
8. Zolotykh, N.Y., Kustikova, V.D., Meerov, I.B.: A review of vehicle detection and tracking methods in video. Vestnik of Lobachevsky University of Nizhni Novgorod **5**(2), 347–357 (2012)
9. Stetsenko, I.V., Stelmakh, O.: Program module for traffic flow intensity determination. Visnyk NTUU "KPI" Inform. Oper. Comput. Sci. **66**, 94–99 (2017)
10. Ronneberger, O., Fischer, P., Brox, T.: U-Net: convolutional networks for biomedical image segmentation. In: Navab, N., Hornegger, J., Wells, W., Frangi, A. (eds.) Medical Image Computing and Computer-Assisted Intervention – MICCAI 2015, Lecture Notes in Computer Science, vol. 9351, pp. 234–241. Springer, Cham (2015)
11. Kingma, D.P., Ba, J.: Adam: a method for stochastic optimization. In: Kingma, D.P., Ba, J. (eds.) 3rd International Conference on Learning Representations (ICLR 2015). Conference Track Proceedings, pp. 1–13. Ithaca, NY: arXiv.org, San Diego (2015)
12. Khizhnyak, A.V., Kuzmenok, M.D.: Recognizing images using convolutional networks. Digital Library of Polotsk State University (2018). http://elib.psu.by:8080/handle/123456789/22222. Accessed 25 Apr 2019

Method for Searching of an Optimal Scenario of Impact in Cognitive Maps During Information Operations Recognition

Oleh Dmytrenko[1], Dmitry Lande[1,2], and Oleh Andriichuk[1,2(✉)]

[1] Institute for Information Recording of National Academy of Sciences
of Ukraine, Kiev, Ukraine
dmytrenko.o@gmail.com, dwlande@gmail.com,
andriichuk@ipri.kiev.ua
[2] National Technical University of Ukraine "Igor Sikorsky Kyiv Polytechnic
Institute", Kiev, Ukraine

Abstract. In this paper, we consider cognitive maps as an additional tool for building a knowledge base of the DSS. Here we present the problem of choosing the optimal scenario of the impact between nodes in the cognitive maps based on of the introduced criteria for the optimality of the impact. Two criteria for the optimality of the impact, which are called the force of impact and the speed of implementation of the scenario, are considered. To obtain a unique solution of the problem, a multi-criterial assessment of the received scenarios using the Pareto principle was applied. Based on the criteria of a force of impact and the speed of implementation of the scenario, the choice of the optimal scenario of impact was justified. The results and advantages of the proposed approach in comparison with the Kosko model are presented. Also we calculate rank distribution of nodes according to the degree of their impact on each other to reveal key and the most influential components of the cognitive map that corresponds some subject domain.

Keywords: Cognitive map · Optimal scenario of impact · Pareto principle · Algorithm of accumulative impact · Force of impact · Rank distribution · Information operation recognition

1 Introduction

In today's world, it is difficult to overestimate the impact of information on people. Recently, the number of information sources has increased significantly and, accordingly, their influence is also increased. Information operations [1] may be one of the negative manifestations of this effect.

During the recognition of information operations [1], decision support systems (DSS) are used to make recommendations. When building knowledge bases of DSSs it often encounters the problem of lack of knowledge for describing a subject domain, which is corresponded to an object of an informational operation. In this case, a cognitive map, which is built automatically based on the textual data that corresponding to the object of the information operation, can be an additional tool for

© Springer Nature Switzerland AG 2020
A. Palagin et al. (Eds.): MODS 2019, AISC 1019, pp. 182–193, 2020.
https://doi.org/10.1007/978-3-030-25741-5_19

building a knowledge base of the DSS. Such a cognitive map is a network of key terms that influence each other. Rank distribution of nodes according to the degree of their impact on each other makes it possible to reveal the key and the most influential components of the subject domain.

Rank distribution is one of the methods of ordering objects either physical or informational. In the case of certain numerical value can be assigned to each object from the collection, the ranking problems become formally trivial, since objects can be ranked by the value [2]. For example, the introduction of weight coefficients, characterizing the power of impact, turned out to be the main direction of development of the cognitive approach for analyzing a situation [3].

A cognitive map is a directed graph in which the edges (and sometimes the nodes) are characterized by weighted factors. A cognitive map, like any graph, is defined by the adjacency matrix W [4], comprised of elements w_{ij} – representing weight values of the edges connecting the corresponding nodes $u_1, u_2, \ldots, u_n$. The nodes of the cognitive map correspond to certain concepts, and edges are the casual (causal-consequential) connections between the corresponding concepts. Weight values are also used to analyze well-structured situations, where the value of the impact in different paths between the two nodes is summed up. However, the difficulty is that, firstly, it is not always clear how to determine such a numerical value, and secondly, such numerical values may be many and not always clear criterion for choosing one of them. In other words, the most complex, poorly formalized part of the problem of ranking is the choice of criterion for which the object is attributed to numerical values (formalization of objects).

In this paper, the value of impact is calculated as follows:

1. In order to calculate the force of impact of one node on another (the impact of u_i on u_j), it is necessary to find all the simple paths that exist between these two nodes. To find all the simple paths between a pair of nodes (u_i, u_j), the algorithm presented in work [5] is used. Each simple path represents a certain scenario of impact $(u_i, u_j)_k$.
2. Having introduced the criteria, the scenario of impact can be considered optimal for: the force and speed of the implementation of the scenario.

The purpose of this paper is to justify a choice the optimal scenario of impact according to the introduced criteria.

2   Methods and Models for Nodes Ranking

In this section a short survey of other methods that can be used for cognitive maps for ranking of nodes according to the degree of their impact on each other makes is presented.

In the impulse method [6], each node in a cognitive map is assigned a value $v_i(t)$ at each moment of discrete time $t = 0, 1, 2, \ldots$. The weight of an edge is positive $(w_{ij} > 0)$ if an increase in the weight of node u_i causes an increase in the weight of node u_j. Conversely, the weight of an edge has a negative value $(w_{ij} < 0)$ if decreasing the weight of node u_i results in a decrease in the weight of node u_j. The weight $w_{ij} = 0$ if nodes u_i and u_j are not related.

The problem is to define the final value of node $v_i(t \to \infty)$, or in some cases the rate of change over time. To define $v_i(t)$ it is necessary to define how the node's value changes depending on its initial value, values of neighboring nodes, and weights of relations.

The basic procedure of cognitive mapping analysis is determined by the rule of the impulse process changing which is described in detail in [6]. According to this rule, the value of each concept $v_i(t)$ changes at the moment of discrete time t ($t = 0, 1, 2,....$) by the following equation:

$$v_i(t+1) = v_i(t) + \sum_{j=1}^{n} w_{ij} p_j(t), t = 0, 1, 2, \ldots \tag{1}$$

where n is the number of nodes in the graph.

An impulse is defined by the following equation:

$$p_j(t) = v_j(t) - v_j(t-1), t > 0. \tag{2}$$

While investigating cognitive maps, values $v_i(0)$, which correspond to the concepts of the directed graph, and the pulse values $p_i(0)$ are defined at the initial moment of time $t = 0$.

In the Kosko model [7, 8] an influence value is calculated as follows: the indirect influence (i.e., the indirect effect) of action I_p of vertex i on vertex j through path P that connects vertex i to vertex j is defined as $I_p = \min_{(k,l) \in E(P)} w_{kl}$, where $E(P)$ is a set of edges along path P and w_{kl} is the weight of edge (k, l) of path P, the value of which is defined in terms of the linguistic variables.

The general influence $Inf_{km}(i,j)$ of vertex i on vertex j is defined as follows: $Inf_{km}(i,j) = \max_{p(i,j)} I_p$, where max is the maximum value along all possible paths from vertex i to vertex j. Thus, I_p defines the weakest link in path P, and $Inf_{km}(i,j)$ defines the strongest influence among the indirect influences I_p.

3   Criteria of Optimality of Impact

Considering each possible simple path from node u_i to the node u_j of cognitive map as a certain scenario of impact $(u_i, u_j)_k$, it is necessity to determine criteria for choosing one of them.

The paper presents two criteria for optimality of impact C_1 and C_2, which are called the force of impact and speed of implementation of the scenario respectively.

The force of impact of node u_i on node u_j is calculated for every path while considering the weights of the edges. The impulse from node u_i is distributed along the path in the direction from u_i to u_j according to rules (a)–(d) [5]:

(a) $u_i \xrightarrow{+} u_k \xrightarrow{-} u_j$

If node u_i has a positive impact on node u_k and node u_k has a negative impact on node u_j, then node u_i is said to increase the negative impact of u_k on u_j. As a result, node u_i is said to have a negative impact on u_j.

(b) $u_i \xrightarrow{-} u_k \xrightarrow{-} u_j$

If node u_i decreases the negative impact of node u_k on u_j, then node u_i is said to have a positive impact on u_j.

(c) $u_i \xrightarrow{+} u_k \xrightarrow{+} u_j$

In this case, u_i has a positive impact on u_j, which increases the positive impact of node u_k on u_j.

(d) $u_i \xrightarrow{-} u_k \xrightarrow{+} u_j$

In this case, node u_i has a negative impact on node u_k and u_k has a positive impact on u_j. In other words, node u_i decreases the positive impact of u_k on u_j. Thus, node u_i has a negative impact on u_j.

The full impact z_{ij} on the node u_j, which is accumulated from the node u_i, is the sum of the partial impacts calculated as subtract between $z_{ij}^k - \tilde{z}_{ij}^k$ in all simple paths from node u_i to node u_j (following to the algorithm for calculating of a mutual impact between nodes in weighted graphs – the algorithm of an accumulative impact, which is presented in [5]) where

$$z_{ij}^k(t+1) = \left(1 + \text{sign}\left(z_{ij}^k(t)\right) * \alpha\left(\left|\frac{z_{ij}^k(t)}{\mu}\right|\right)\right) * w\left(q_t^k, q_{t+1}^k\right) \qquad (3)$$

$$\tilde{z}_{ij}^k(r+1) = \left(1 + \text{sign}\left(\tilde{z}_{ij}^k(r)\right) * \alpha\left(\left|\frac{\tilde{z}_{ij}^k(r)}{\mu}\right|\right)\right) * w\left(q_r^k, q_{r+1}^k\right) \qquad (4)$$

where sign is a signum function;
   q_t^k – the sequence of nodes included in the k-th path ($q_0 = u_i$, $q_{m-1} = u_j$);
   $t = 0, 1,\ldots,m-2$, a $r = 1,\ldots,m-2$, (m is the number of nodes included to the k-th path).
   Here, the initial conditions are: $z_{ij}^k(0) = 0, \tilde{z}_{ij}^k(1) = 0$.

$$\mu = \max\left|w_{ij}\right|, \qquad (5)$$

where $i = 0, 1,\ldots,n, j = 0, 1,\ldots,n$ (n is the dimension of the cognitive map).
   The impact of a node u_j on a node u_j is called "the strongest", if the partial impact on the final node is characterized by the greatest of absolute magnitude of an impact among all of the partial impacts on all simple paths between two nodes u_j and u_j.
   The impact of the node u_j on node u_j is considered to be "the fastest in realization", if it is carried out in the shortest path. The speed of the implementation of the k-th scenario is determined by the number of edges ($m-1$) connecting the nodes u_j and u_j in k-th path (where m is the number of nodes included in the k-th path).
   The introduced criteria of C_1 and C_2 are almost equivalent in terms of priority, thereby if one get several different optimal scenarios $(u_i, u_j)_k$ of the impact of the node,

one cannot just select neither of them. Therefore, the fundamental complexity of choice in multi-criteria problems consists in impossibility of determining the optimal scenario a priori. So, there is a need to compare alternatives to all criteria.

Let us consider X as a set of possible scenarios (alternatives) $(u_i, u_j)_k$ of the impact of the node u_i on u_j. The minimum number of elements included in the set X is two (to be able to make a choice). There is no limit on the number of possible scenarios: the number of elements of the set can be both finite and infinite. It is worth noting that sometimes a choice of not one, but an entire set of decisions is made, which is a subset of a set of possible solutions X. In this paper, it is necessary to justify the choice of the optimal scenario of impact according to the introduced criteria C_1 and C_2. Then $C(X)$ is a set of selected scenario. It is a solution of the problem of choice and it can be any subset of the set of possible scenarios X. Thus, solving the problem of choice means to find a subset of $C(X)$, $C(X) \subset X$.

In the case where a plurality of selected scenarios does not contain any element, the choice does not occur, due to the fact that no solution has been selected. That is, in order to make the choice, it is necessary that the set $C(X)$ contains at least one element.

There are various methods for solving multi-criteria problem [9]. In order to obtain a unified solution to the problem posed in this paper, a multi-criteria assessment of the scenarios obtained according to the Pareto principle [10–12] is used.

Pareto's approach is as follows: the alternative is "the best" than the alternative for Pareto $(x \succ y)$, if alternative x alternatives are rated "no worse" than alternatives y, and at least one alternative x is "the best" than alternatives y:

$$\forall_i C_i(x) \geq C_i(y) \text{ and } \exists_j : C_j(x) > C_j(y) \tag{6}$$

where $C(X)$ is a function of choice $(C(X) \subset X)$.

The resulting set of solutions is called pareto-optimal.

4   Method for Searching of an Optimal Scenario of Impact

Let us set of possible vectors X consist of a finite number of elements N and has the form $X = \{x^{(1)}, x^{(2)}, \ldots, x^{(N)}\}$.

In order to construct it on the basis of the definition of the Pareto set, it is necessary to compare each vector $x^{(i)} \in X$ with any other vector $x^{(j)} \in X$. Thereby, a step-by-step comparison of scenarios (corresponding columns of the table) based on the principle of "no less" ("no more") according to all criteria is performed. Namely: if the i-th scenario is larger (at minimization) or smaller (at maximization) of j-th scenario by at least one criterion, then this scenario is no longer taken into account. But if at least one i-th scenario criterion is less (at minimization) or larger (at maximization) for j-th scenario, with one or more other criteria, it is greater (at minimization) or smaller (at maximization), then both scenarios are taken into account.

It must be pointed out that it is convenient to use a table whose rows are criteria C_1 and C_2 (a force of impact and ease of implementation of the scenario, respectively), and the columns are the number of a scenario $(u_i, u_j)_k$ (the numbers of simple paths connecting the nodes u_i and u_j) for comparison alternatives.

Thus, columns of a table form a set of possible vectors (possible scenarios), which consist of two elements - the values of the criteria. The result of a staged comparison is the set $C(X)$ of such non-extracted vectors forms the Pareto set. But often this is the case, and as already mentioned above, the Pareto set may contain more than one element. These are scenarios that cannot be compared according to the Pareto principle. In the general case, when the Pareto set contains more than one element, in order to determine the optimal scenario of impact in this paper, the following algorithm is proposed:

(a) Firstly, the least common multiple (LCM) of the criterion C_2 for all values of the Pareto set is determined. Considering C_2 as time the corresponding scenario is implemented for, then the LCM of all values is the least time for which the integer number of each of the scenarios included in the Pareto set is realized. Thereby, at the same time $LCM(c_2^{(1)}, \ldots, c_2^{(d)})$, the number of realizations of the various scenarios included in the Pareto set are different accordingly $\{a^{(1)}, a^{(2)}, \ldots, a^{(d)}\}$:

$$a^{(k)} = \frac{LCM(c_2^{(1)}, \ldots, c_2^{(d)})}{c_2^{(k)}} \qquad (7)$$

where $c_2^{(k)}$ – value of the criterion C_2 for k-th scenario;
  d – the number of elements included in the Pareto set.

(b) Next, for each of the scenarios included in the Pareto set, the values of their assessments by the criterion C_1 $\left\{c_1^{(1)}, c_1^{(2)}, \ldots, c_1^{(d)}\right\}$ are multiplied by the corresponding value $\{a^{(1)}, a^{(2)}, \ldots, a^{(d)}\}$. That is, it determines what will be the overall impact of the node u_i on node u_j the time $LCM(c_2^{(1)}, \ldots, c_2^{(d)})$ of the k-th scenario.

(c) In order to determine the optimal scenario of impact, it is necessary to find the highest value of the multiplication $c_1^{(k)} \cdot a^{(k)}$ defined in step (b):

$$\max_k = c_1^{(k)} \cdot a^{(k)} \qquad (8)$$

where $k = 1, .., d$.

That is, the number k to which the largest multiplication $c_1^{(k)} \cdot a^{(k)}$ corresponds is the number of the optimal scenarios of impact. As a result of the justification of the choice of the optimal criteria C_1 and C_2 the impact scenario for each pair of nodes (u_i, u_j) of a weighted graph, we can construct a matrix Z which consists of elements z_{ij} and a matrix T which consists of elements t_{ij}.

Definition 1: The full impact z_{ij} is the partial impact of the node u_i on the node u_j, which is accumulated in accordance with the optimal scenario of impact (i.e., the value of the criterion C_1 of the optimal scenario of impact). If u_j is unavailable from node u_i, then $z_{ij} = 0$.

Definition 2: The full time t_{ij} is the time it takes to implement the optimal scenario of impact node u_i on node u_j (i.e., the value of the criterion C_2 of the optimal scenario of impact). If u_j is unreachable from the node u_i, then $t_{ij} = 0$.

In order to determine what the impact of each of the nodes at $t \to \infty$, must be fulfilled as follows. Firstly z_{ij}^1 needed to be defined z_{ij}^1 - the impact of each node at $t = 1$. Taking into account the time required to implement each of the scenarios for the impact matrix Z by dividing each of its non-zero elements z_{ij} $(z_{ij} \neq 0)$ into the corresponding element t_{ij} of the matrix T, a matrix Z_1 is obtained, the elements of which are:

$$z_{ij}^1 = \begin{cases} \frac{z_{ij}}{t_{ij}}, \ z_{ij} \neq 0 \\ 0, \ z_{ij} = 0 \end{cases}. \tag{9}$$

Next, the impact of each node in time t is calculated as $z_{ij}^{t+1} = z_{ij}^t + z_{ij}^1$. At each step of $t = 2, 3, 4 \ldots$, the process of normalization is carried out:

$$z_{ij}^t = \frac{z_{ij}^t}{\sum\limits_{k=1}^{n} \sum\limits_{l=1}^{n} z_{kl}^t}. \tag{10}$$

5   Example

The work [12] considered the weighted directed graph shown in Fig. 1.

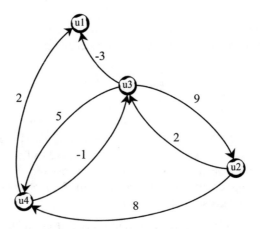

Fig. 1. Weighted directed graph.

The weighted directed graph, presented in Fig. 1, corresponded to cognitive map is defined by the adjacency matrix:

$$W = \begin{pmatrix} 0 & 0 & 0 & 0 \\ 0 & 0 & 2 & 8 \\ -3 & 9 & 0 & 5 \\ 2 & 0 & -1 & 0 \end{pmatrix} \tag{11}$$

Table 1 demonstrates an example of assessment of impact scenario of nodes u_3 on nodes u_4 by criteria C_1 and C_2:

Table 1. Example of assessment of impact scenario of nodes

$(u_3, u_4)_k$	Simple path from u_1 to u_4	C_1	C_2
1	$u_3 \xrightarrow{5} u_4$	6.92	2
2	$u_3 \xrightarrow{9} u_2 \xrightarrow{8} u_4$	5	1

Table 2 shows the Pareto table to find the optimal criteria C_1 and C_2 scenario of the node u_3 impact on node u_4 for the cognitive map, which is shown in Fig. 1.

Table 2. Pareto table to find the optimal criteria

$(u_3, u_4)_k$	1	2
C_1	6.92	5
C_2	2	1

In this case, the Pareto set consists of two non-comparable vectors (two scenarios 1 and 2), among which it is impossible to determine uniquely optimal by criteria C_1 and C_2 (scenario 1 is the optimal by criterion C_1, and scenario 2 is by C_2 one). Therefore, for the final solution of the problem of choosing the optimal scenario of impact, it is necessary to determine the alternative to the optimal solution for a particular practical problem.

According to the method for searching of an optimal scenario of impact, which is proposed in this paper, when the Pareto set contains more than one element, it is first necessary to find LCM of values of the criterion C_2 values for all elements of the Pareto set. For the Pareto set constructed from the set of alternatives presented in Table 2, the least time for which the integer number of each of the scenarios included in this Pareto set is equal:

$$LCM(2, 1) = 2.$$

The number of implementations of the first and second scenarios will be equal respectively

$$a^{(1)} = \frac{LCM(2,1)}{2} = \frac{2}{2} = 1,$$

$$a^{(2)} = \frac{LCM(2,1)}{1} = 2.$$

Over time equal to $LCM(2,1) = 2$, the overall impact of the node u_3 on node u_4 the 1st scenario $(u_3, u_4)_1$ is:

$$c_1^{(1)} \cdot a^{(1)} = 6.92 \cdot 1 = 6.92$$

In the 2nd scenario

$$c_1^{(2)} \cdot a^{(2)} = 5 \cdot 2 = 10.$$

$$\max(c_1^{(1)} \cdot a^{(1)}, c_1^{(2)} \cdot a^{(2)}) = \max(6.92, 10) = 10$$

Therefore, for this example (Table 2), scenario number 2 $(u_3, u_4)_2$, in accordance with the proposed method, is optimal.

The Pareto table for other pairs of nodes (u_i, u_j) $(i, j = 1, 2, 3, 4)$ with more than one scenario of impact is given in Table 3.

Table 3. Pareto table for other pairs of nodes

$(u_2, u_1)_k$	1	2	3	4
C_1	-1.08	0.41	1.66	0.22
C_2	2	3	2	3
$(u_2, u_3)_k$	1	2		
C_1	2	-0.83		
C_2	1	2		
$(u_2, u_4)_k$	1	2		
C_1	8	1.79		
C_2	1	2		
$(u_3, u_1)_k$	1	2	3	
C_1	-3	0.27	1.34	
C_2	1	3	2	
$(u_4, u_1)_k$	1	2		
C_1	0.59	2		
C_2	2	1		

For all possible scenarios presented in Table 3 for pairs (u_2, u_1), the Pareto set will consist of one element $(u_2, u_1)_3$ - from scenario 3. For a pair (u_2, u_3), the Pareto set consists of the element $(u_2, u_3)_1$ - scenario 1. For $-(u_2, u_4) - (u_2, u_4)_1$, for $(u_3, u_1) - (u_3, u_1)_1$- and for the pair - $(u_4, u_1) - (u_4, u_1)_2$.

After choosing the best scenario of impact by the introduced criteria C_1 and C_2 for each pair of nodes (u_i, u_j) $(i, j = 1, 2, 3, 4)$ of the weighted directed graph represented in Fig. 1, we can construct an influence matrix Z which consists of elements z_{ij} (see Definition 1) and a matrix T which consists of elements (see Definition 2):

$$Z = \begin{pmatrix} 0 & 0 & 0 & 0 \\ 1.66 & 0 & 2 & 8 \\ -3 & 9 & 0 & 5 \\ 2 & -1.79 & -1 & 0 \end{pmatrix}$$

$$T = \begin{pmatrix} 0 & 0 & 0 & 0 \\ 2 & 0 & 1 & 1 \\ 1 & 1 & 0 & 1 \\ 1 & 2 & 1 & 0 \end{pmatrix}$$

Taking into account the process of normalization at each step with $t \to \infty$, the impact of each of the nodes to other for the influence matrix Z is represented in the form of the influence matrix:

$$Z_t = \begin{pmatrix} 0 & 0 & 0 & 0 \\ 0.038 & 0 & 0.091 & 0.365 \\ -0.137 & 0.41 & 0 & 0.228 \\ 0.091 & -0.04 & -0.046 & 0 \end{pmatrix} \tag{12}$$

The full impact Inf_{pm} of each node for the influence matrix Z_t is determined by the rule:

$$Inf_{pm}^i = \sum_{j=1}^{n} |z_{ij}^t| \tag{13}$$

where n is the number of nodes of a cognitive map; pm is a short for "the Pareto method".

The full impact Inf_{pm} of each node and its rank distribution for the influence matrix (12), according to (13), are presented in Table 4.

Comparing the results of using the method for searching of an optimal scenario of impact with the results provided by the Kosko model for the adjacency matrix (11), it can be notice (Table 5) that the rank distribution of nodes by degree of impact, as a result of applying of each method, is remained the same.

In Table 5 km is a short for "the Kosko model".

Table 4. Rank distribution of nodes

Node (№)	Inf_{pm}
3	0.775
2	0.494
4	0.178
1	0

Table 5. Rank distribution of nodes according to Kosko model and proposed method

Node (#)	Inf_{km}	Node (#)	Inf_{pm}
3	20	3	0.775
2	12	2	0.494
4	4	4	0.178
1	0	1	0

6   Conclusions

Consequently, the multi-criteria choice problem was considered in the paper. Based on the criteria of a force of impact and speed of the implementation of the scenario, the choice of the optimal scenario of impact was justified. As the result, scenario of impact of node 3 on node 4, which has the number 2, in accordance with the proposed method, is optimal for the considered example. Also the choice of the optimal scenario of impact was justified for other pairs of nodes in the presented cognitive map. A comparison of the results of applying the method for searching of an optimal scenario of impact according to the introduced criteria, with the results which are obtained with applying the Kosko model was fulfilled. It was established that the rank distribution of nodes by degree of impact, as a result of applying of each method, is remained the same.

Using the results of these calculation, decision makers can develop strategic and tactical steps to counter-act the information operation, evaluate the operation's efficiency.

Acknowledgment. This study is funded by the NATO SPS Project CyRADARS (Cyber Rapid Analysis for Defense Awareness of Real-time Situation), Project SPS G5286.

References

1. Dodonov, A.G., Lande, D.V., Tsyganok, V.V., Andriichuk, O.V., Kadenko, S.V., Graivoronskaya, A.N.: Information Operations Recognition: From Nonlinear Analysis to Decision-Making. E-preprint ArXiv: https://arxiv.org/abs/1901.10876
2. Management Association, Information Resources (ed.): Social Media and Networking: Concepts, Methodologies, Tools, and Applications: Concepts, Methodologies, Tools, and Applications. GI Global, pp. 2015–2298
3. Axelrod, R.: Structure of Decision: The Cognitive Maps of Political Elites, p. 422. Princeton University Press, Princeton (2015)

4. Stea, D.: Image and Environment: Cognitive Mapping and Spatial Behavior, p. 439. Routledge (2017)
5. Dmytrenko, O.O., Lande, D.V.: The Algorithm of Accumulated Mutual Influence of the Nodes in Semantic Networks. E-preprint ArXiv (2018). arXiv:1804.07251
6. Roberts, F.: Discrete Mathematical Models with Applications to Social, Biological, and Environmental Problems. Rutgers University, Prentice-Hall Inc., New Jersey (1976)
7. Kosko, B.: Fuzzy cognitive maps. Int. J. Man Mach. Stud. **24**, 65–75 (1986)
8. Kosko, B.: Fuzzy Thinking. Hyperion, New York (1993)
9. Ehrgott, M.: Approximative solution methods for multiobjective combinatorial optimization. Sociedad de Estadistica e Investigacion Operativa **12**(1), 1–63 (2004)
10. Godfrey, P., Shipley, R., Gryz, J.: Algorithms and analyses for maximal vector computation. VLDB J. **16**, 5–28 (2006)
11. He, Z., Yen, G.G., Zhang, J.: Fuzzy-based Pareto optimality for many-objective evolutionary algorithms. Trans. Evol. Comput. **18**, 269–285 (2014)
12. Köppen, M.K., Garcia, R.V., Nickolay, B.: Fuzzy-Pareto-dominance and its application in evolutionary multiobjective optimization. In: Proceedings of International Conference on Evolutionary Multi-Criteria Optimization, pp. 399–412 (2005)

Mathematical Modeling and Simulation of Special Purpose Equipment Samples

Model for Intercepting Targets by the Unmanned Aerial Vehicle

D. N. Kritsky$^{(\boxtimes)}$ (ID), V. M. Ovsiannik (ID), O. K. Pogudina (ID),
V. V. Shevel (ID), and E. A. Druzhinin (ID)

National Aerospace University "KhAI", Chkalova 17, Kharkiv 61070, Ukraine
d.krickiy@khai.edu

Abstract. The system for simulating flight control according to the characteristics of the unmanned aerial vehicle (UAV), as well as the flight paths of the UAV target, is proposed. The overview of the search for the optimal route for the UAV methods is performed. The following tasks were solved: development of the simulation model of the flight of a drone-target and an intercept drone; development of the subsystem of constructing the flight trajectory of the drone-target and building the optimal trajectory of the drone-interceptor; demonstration of the process of intercepting the goal by an intercept drone in real and virtual time, considering its geometric and aerodynamic characteristics.

Keywords: Unmanned aerial vehicles · Drone-interceptor · Drone-target · Optimal trajectory

1 Introduction

Recently, UAV are increasingly replacing standard manned aircraft from various fields of combat use of aircraft, due to a number of conditions [1]: higher survivability of the UAV (due to its lower visibility in the radar, infrared, optical and acoustic ranges); the longest flight duration is calculated around the clock today; less likely to be detected and destroyed by enemy antiaircraft defenses, compared to manned aircraft; the ability to carry out controlled safe flight at extremely low altitudes, sometimes even inaccessible to manned aircraft; the possibility of standing a high alert (operational) readiness for almost an unlimited time, as well as a significantly lower cost of development, mass production and military operation of the UAV and the training of operators of the ground or other command posts (CP).

Modern professional and semi-professional UAV are equipped with various kinds of cameras on board, conduct air patrols along the pipelines, calculate the area of ignition during fires, and have an environmental monitoring [2]. Currently, this set of tasks requires less time to perform, fewer forces, and it is also safe to say that in the future UAV will be used on an ever-larger scale.

One of the advantages that UAV have compared with a manned aircraft is the independence of the maximum flight time from the physiological capabilities of the flight crew. This is a significant advantage in the context of operational and strategic requirements in commensurate with the concepts of "Global strike" and "Global

© Springer Nature Switzerland AG 2020
A. Palagin et al. (Eds.): MODS 2019, AISC 1019, pp. 197–206, 2020.
https://doi.org/10.1007/978-3-030-25741-5_20

sustainable strike" [2]. As an example of the influence of the flight duration factor, can be considered this situation. For a hypothetical combat area measuring 192×192 miles, assuming the above requirement, there must be attack aircraft carrying weapons within 32 miles of any point in the area (five-minute time response for guaranteed destruction of mobile targets), which requires continuous presence in the area at least nine carriers defeat. To this should be added the conditions of basing restrictions, from land or sea bases, with a typical distance of about 1500 miles from the center of the combat area [3].

The key problem in the UAV design is the search for structural compromises between the dimensions of the UAV, combat survival, the size of the ammunition, cost, which determines the number of groups in conditions of limited assignations. The upper level of flight duration from the Global Hawk UAV experience, considering scientific and technological progress, can be several times higher than the level reached in 36 h for this UAV [4].

It should be noted that for combat UAVs, the required duration of stay in the combat area should be determined considering the intensity of spending weapons, ammunition on board, and the levels of its survival. The optimal ratio of fuel reserves and weapons ammunition depends on the predicted conditions of combat use - the intensity of hostilities, and various technical solutions can be used in the process of mission for its operational management, for example, having a modular weapon compartment with the ability to place both fuel and weapons [5].

Cost is a significant limitation on the dimension of the UAV. For conditions of the co-use with manned strike aircraft, the specified UAV accounting parameters, including cost, survival and combat effectiveness, should be determined by complex performance indicators with the search for the optimal composition of the aviation group with manned and unmanned strike systems, and the rational distribution of the share of combat missions between them [6].

The most obvious and effective way to counter the UAV is to identify such equipment with subsequent destruction. To solve this problem can be used as existing samples of military equipment, modified accordingly, as well as new systems [7].

One of the main issues in the destruction of enemy equipment is its identification, followed by maintenance. Modern types of anti-aircraft weapons systems of most types include radar warning station with various characteristics. The probability of detecting an air target depends on some parameters, primarily on its radar cross-section (RCS). Comparatively large UAVs are distinguished by greater RCS, which facilitates their detection. In case of compact devices, including those that was built with extensive use of plastics, the RCS decreases, and the task of detection is seriously complicated [8].

However, during creating promising air defense tools, measures are taken to improve the detection efficiency of targets. This development leads to the expansion of the RCS ranges and target velocities at which it can be detected and taken for tracking [9]. After identifying a potentially dangerous target, it should be identified and determine which object entered the airspace [10]. The correct solution of such a task will make it possible to determine the need for an attack, as well as to establish the characteristics of the target necessary for choosing the right weapon. In some cases, the correct choice of means of destruction can be associated not only with the extra consumption of inappropriate ammunition, but also with the negative consequences of a tactical nature. After successful detection and recognition of enemy equipment, the air defense system must carry out the attack and destroy it. For this should be used weapons that match the type of target detected. For example, large UAVs for recon or attack purposes should be hit with anti-aircraft missiles. In the case of low-altitude and low-speed light-class vehicles, it makes sense to use stem armament with appropriate ammunition. In particular, artillery systems with controlled remote detonation have great potential in the fight against UAVs [4].

The purpose of this publication is to develop algorithms and models of an UAV.

2 Interceptor Flight Path Formation Algorithm

The flowchart (Fig. 1) and method of interceptor flight path formation algorithm were developed.

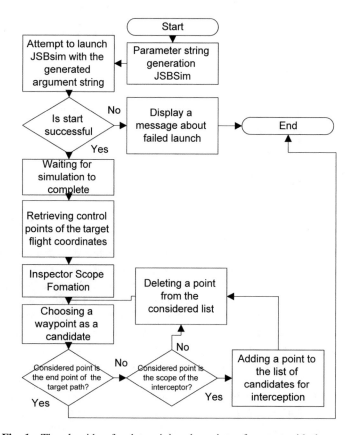

Fig. 1. The algorithm for determining the points of contact with the target

In the process of determining the trajectory of the interceptor, a guidance method using the "chase curve" principle was applied. This method was chosen because of its relatively simple implementation, as well as because of the possibility of using it when working not only in the guidance system of self-guided missiles, but also with UAV [11].

The method of pointing along a chase curve resembles the pursuit of a dog after a hare, therefore another name of this method can be found in the literature - pointing along a chasing curve or "dog curve" [12].

In this method, two possible cases should be considered: the achievement of a goal that is moving away (when passing courses, Fig. 2a) and the achievement of a goal that approaches (Fig. 2b).

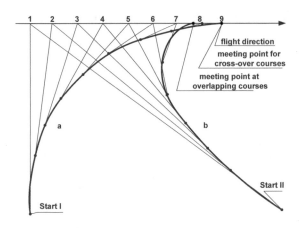

Fig. 2. Curves pursuing for passing-intersecting and counter-intersecting courses

In the first case (Fig. 2a), if the aircraft has a sufficient range and speed greater than the speed of the target, it can hit the target.

In the second case (Fig. 2b), as the aircraft approaches the target, the speed of rotation sharply increases. Such a load is not able to withstand the rocket body, and, in this case, it would simply break. But in reality, the control force created by the rudder can only grow to a certain value. Consequently, there may come a time when the steering wheel of the aircraft deviates to the stop, but the value that arises from this maximum control force will be insufficient for the necessary change of the direction.

From this point on, the aircraft will begin to move in a circle of minimum radius, which corresponds to the limiting control force. Guidance will stop, as the aircraft will not have time to develop on the target. After some time, the goal will leave the field of the coordinator's view, after which the guidance becomes impossible [12].

To derive the equation of the line, choose a coordinate system in which the abscissa axis passes through the initial position of the points P and A, the point A is at the origin of the xAy coordinate system. The ratio of constant velocities of points is denoted by k.

If we assume that for an infinitely small period of time the point P has covered the distance dS, and the point A - distance dS_1, then, according to the above condition, we obtain the relation $dS = k\,dS_1$, or

$$\sqrt{dx^2 + dy^2} = k\sqrt{d\xi^2 + d\eta^2},$$

Further it is necessary to express $d\xi$ and $d\eta$ through x, y and their differentials. Provided, the coordinates of the point P must satisfy the equation of the tangent to the desired curve, that is, $\eta - y = \frac{dy}{dx}(\xi - x)$.

Adding to this equation the equation $F(\xi, \eta)$ for the "evader" motion path specified by the condition, can be determined from the resulting system of equations ξ and η. After substitution of these values into the differential equation, it will be written as

$$\Phi\left(x, y, \frac{dy}{dx}, \frac{d^2y}{dx^2}\right) = 0.$$

Integration constants can be found from initial conditions ($y = 0$; $y' = 0$ when $x = 0$).

In the general case, for an arbitrarily given curve $F(\xi, \eta)$, it is rather difficult to find a solution to the resulting equation. The task is greatly simplified if we consider the simple case when the trajectory of the "evader" is straight.

Consider the case of A_0 (0, 0), P_0(0, 1) when the "evader" moves along the x axis and when $k > 0$. At any given time, the "evader" is always on a tangent to the curve of the "pursuer's" motion path, so $\frac{dy}{dx} = \frac{-y}{a-x}$.

Based on what let's write the differential equation $y + y'(a-x) = 0$, where $y > 0$.

From the condition $a = V \bullet t$ floats $\frac{y}{y'} + Vt = x$, after differentiation by time $\dot{y} = y' \bullet \dot{x}$ and $\ddot{y} = y'' \bullet \dot{x}$, on the basis of which find out $\dot{x} = \frac{dx}{dt} = \frac{V \cdot y'^2}{y \cdot y''}$

Let's write the expression to determine the length of the curve

$$l = Wt = k\int_0^x \sqrt{1 + (y')^2}\,dx$$

From the expressions $dx^2 + dy^2 = W^2dt^2$ and $w^2 = \frac{dx^2}{dt^2} + \frac{dy^2}{dt^2} = \dot{x}^2 + (y' \cdot \dot{x})^2$ leaks out $\dot{x} = \frac{w}{\sqrt{1 + y'^2}}$

Similarly, differentiation by y carried out $y'' - k \cdot \frac{y'^2}{y} \times \sqrt{1 + y'^2} = 0$

Substitution Solution $u = x' = \frac{1}{y'}$, $y'' = \frac{-1}{u^3}\frac{du}{dx}$, with separation of variables leads to $\frac{-du}{\sqrt{1 + u^2}} = k \cdot \frac{dy}{y}$ after integration we get arsinh $u = k \cdot \ln y + C$.

Further, after using the formal definition sinh with $C_1 = e^C$, we obtain $x' = \frac{dx}{dy} = \frac{1}{2}\left[(C_1 \cdot y)^k - (C_1 \cdot y)^{-k}\right]$

Integrate once again with the definition of continuous integration C_2. From the initial conditions $\left.\frac{dx}{dy}\right|_{y=1} = 0$ leaks out $C_1 = 1$, and also $x|_{y=1} = 0$ the result is

$$C_2 = \frac{k}{1-k^2} \text{ or } x(y) = \frac{1}{2}\left(\frac{y^{(1+k)}}{(1+k)} - \left\{\begin{matrix} \frac{y^{(1-k)}}{(1-k)} \\ \ln|y| \end{matrix}\right\}\right) + \left\{\begin{matrix} \frac{k}{(1-k^2)} \\ -1/4 \end{matrix}\right\}\left\{\begin{matrix} k \neq 1 \\ k = 1 \end{matrix}\right.$$

Based on these equations it was possible to obtain the equations that were given above.

3   Software Architecture

"InterceptionUAVSimulation" software system is a software system that consists of a user interface module—an interface implemented using the XAML markup language. It integrates the JSBSim flight dynamics modeling system, the WPF. Media animated data modeling system, the XSeed data visualization module, and the UAV flight path design module, implemented as a set of classes written in C #, into a single complex. This set of modules and subsystems together constitute a single system for solving the problem of designing and developing simulation models of a UAV target along a given route, taking into account the relief and the UAV interceptor, based on tactical and technical characteristics of the target and its trajectory.

Figure 3 show the general file architecture of the "InterceptionUAVSimulation system".

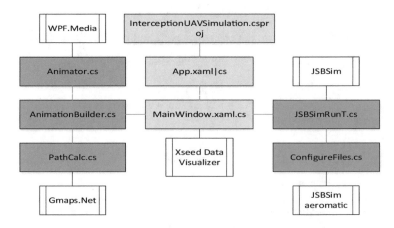

Fig. 3. Software file structure

The darker color in Fig. 3 shows files that contain the main logic classes of the software system that perform certain tasks in the simulation process. Description of the main modules of the system used to achieve the objectives of this work is in below.

The Animator.cs file is a class containing the logic for designing animated GUI objects of software. The main purpose of this module is the formation of graphical representations of the target interception model of the drone.

The PathCalc.cs file contains a class that implements the logic for conducting geolocation calculations and scaling UAV flight routes. There are temporary calculations of UAV flight routes, scaling the speed of the animated model during the demonstration.

The AnimationBuilder.cs file describes a class that contains the logic of matching data obtained during the operation of the Animation and PathCalc modules, after which the final animated model is designed and provided to the system for its further demonstration to the user.

The ConfigureFiles.cs file contains class definitions that contain the logic for designing UAV configuration files, creating their XML markup, creating the directory structure, that a necessary for the correct operation of the JSBSim system during flight simulation.

The JSBSimRunT.cs file describes the main class for working with the JSBSim system. This class is used to build queries to the JSBSim system, to verify the correctness of the file structure of the UAV configuration. This module also implements a system for notifying the user about errors in the system in case of inconsistencies in the location of configuration files UAV representation.

After processing the received data, the XSeed subsystem provides InterceptionUAVSimulation with a set of graphical representations of the UAV telemetry readings during the flight. In turn, GMaps.Net, on the basis of the data obtained, forms a set of interface elements for visualizing the flight process, after which it transfers them to WPF.Media to create an animated representation of the flight process of the UAV.

4 Deploying the Software System

In the main window of the program, the user needs to fill in information about the main design features of the UAV target, such as wing area, fuselage length, maximum lifting weight, relative coordinates of the center of gravity, relative coordinates of the point of view that are needed to visualize the flight of the aircraft, the location and number of engines, the area of horizontal and vertical tail, as well as the name of the projected aircraft. In addition to the basic characteristics of the aircraft, user must enter the characteristics of the engine installed on the aircraft.

Next, the user goes to the window of registration files configuration UAV interceptor. After filling in the configuration files of the target and the interceptor, a transition takes place to the map operation window (Fig. 4).

On this form, the user must specify the targets and coordinates of the location of the launch point of the interceptor using double clicks on the map of control points of the proposed route. Switching between the target route setting modes and the interceptor launch point is performed using the appropriate controls. If an error occurs during the task of markers on the map, it is possible to reset the current location of the markers by clicking on the "Clear target route" button.

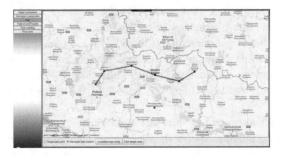

Fig. 4. Map processing form after completing route planning

After completing the map layout (Fig. 4), click the "Complete map config" button, after which the system will launch the flight scenario design module for the target.

After the map layout has been completed and the flight script file has been designed, a transition is made to the launch form of the design of a simulation flight model of the target and interceptor (Fig. 5). On this form, the user can see the directories in which the configuration files of the UAV-target and the interceptor are located, in case of extreme need open them to make changes.

Fig. 5. JSBSim system after successful completion of the simulation

Also on this form there is a logging area, in which messages about events that have arisen in the course of aircraft simulation are displayed. After reviewing the received configuration files, user should click the "Start config flight dynamic models" button, after which the JSBSim system will be launched alternately for the target and the interceptor.

In the intervals between JSBSim launches, the system will analyze and extract the geolocation data of the target during the flight. This data set is used by the system to calculate the interceptor's flight path as calculated by the homing method along the pursuit curve.

Upon successful completion of the simulation, a corresponding message will be displayed in the logging area (Fig. 5) and a flight report file will be created on the disk for the target and the interceptor.

After the simulation of the interception process is completed, data is read from the target and interceptor files, parsed for further visualization using an animated model and graphical reports, and the transition to the form demonstrating the process of intercepting the target. First, on this form, in the map location area, the obtained flight trajectories are shown. In the lower part of the form, there are controls for the playback speed of the animation model and the start/pause of the demonstration. In the process of modeling, the user is provided with a simplified visual process of pursuing, neutralizing and returning to the interceptor base. The interceptor is marked in red on the form, and the target is in green. During the demonstration, the user is given the opportunity to zoom in time using the control element in the lower left area of the form, by default, the visualization takes place in real time. After neutralization, the target is painted black, the rendering stops and a message about the success of the interception process is displayed, after which the interceptor starts to return to the launch base.

When switching to the form of graphical reports (Fig. 6), the user can view the metric readings obtained during the simulation by selecting the appropriate indicator and its membership (goal or interceptor).

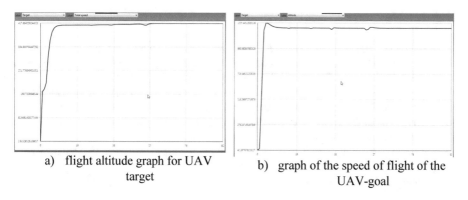

   a) flight altitude graph for UAV target
   b) graph of the speed of flight of the UAV-goal

Fig. 6. Graphical reports

5   Conclusions

In the course of the project, the task of designing and developing a simulation model of the process of interception of the UAV-target by the UAV-interceptor was solved.

First, the analysis of the subject area was carried out, various systems and methods of countering the UAV were considered, possible ways of neutralizing the UAV were studied. Further, the decomposition of the project tasks into subtasks was performed and algorithms for solving these subtasks were compiled.

Analyzing the velocity plots, as well as the results of modeling the construction of trajectories for target intercepting, we can conclude that the developed algorithm allows us to construct an intercept trajectory in such a way that the interception process itself is carried out in the shortest possible time.

Next, a software system was created that builds the flight path of the target, calculates the flight path of the target, calculates the interceptor's flight curve using the chase curve method, and also built an animated model of the target interception process.

The developed system uses the JSBSim flight dynamics simulation program, the XSeed graphic report generation module that is used in the process of generating graphs of changing UAV characteristics during the flight, the geographic points targeting subsystem on the world map GMaps.Net, as well as the main software product WPF application that connects the entire set of modules into a single system.

The developed software system will help researchers and developers of UAVs to determine the optimal geometric and aerodynamic parameters of the aircraft, necessary to achieve the goals and objectives, as well as to determine the possible trajectory of interception of the target.

The practical value of the obtained results is that the work, that have been done, will help in the development of algorithms for automatic control of the UAV-interceptor.

References

1. Kritskiy, D., Karatanov, A., Koba, S., Druzhinin, E.: Increasing the reliability of drones due to the use of quaternions in motion. In: 2018 IEEE 9th International Conference on Dependable Systems, Services and Technologies (DESSERT) (2018)
2. How to destroy a drone. http://www.popmech.ru/weapon/15671-kak-ubit-bespilotnik/
3. How to shoot down a drone. http://blog.i.ua/user/1767616/1903344/, (in Russian)
4. Organization of the system to combat small-sized UAVs. http://arsenal-otechestva.ru/article/389-antidrone, (in Russian)
5. Detection and counteraction to drones. http://robotrends.ru/robopedia/obnaruzhenie-i-protivodyaystvie-bespilotnikam, (in Russian)
6. Modern drones in detail. http://voprosik.net/sovremennye-bespilotniki-podrobno/. (in Russian)
7. Old and new ways of dealing with unmanned vehicles. https://topwar.ru/37629-starye-i-novye-sposoby-borby-s-bespilotnymi-apparatami.html, (in Russian)
8. UAV battle. https://defence-ua.com/index.php/statti/1235-borotba-z-bpla-2, (in Ukranian)
9. Ways of dealing with unmanned aerial vehicles. http://army-news.ru/2013/12/sposoby-borby-s-bespilotnymi-apparatami/, (in Russian)
10. The system of fast detection and suppression of UAVs. http://www.secnews.ru/foreign/22180.htm#axzz4Xin3eaWY, (in Russian)
11. General principles of homing. http://samonavedenie-raket.ru/obshchie-printsipy-samonave-deniya, (in Russian)
12. The method of guidance for the chase curve. http://samonavedenie-raket.ru/metody-navedeniya/metod-navedeniya-krivoi-pogoni, (in Russian)

Mathematical Modeling and Simulation
of Systems in Project Management

Modeling of Empathy, Emotional Intelligence and Transformational Leadership to the Project Success

Bushuyev Sergey$^{(\boxtimes)}$ ⓘ, Kozyr Boris ⓘ, and Rusan Nadiia

Kiev National University of Construction and Architecture, Kiev, Ukraine
SBushuyev@ukr.net

Abstract. The emotional intelligence that many speak in recent times is not only the ability to control their feelings, but also the ability to understand other people. Manager is simulating it project manager performance more extensive features, which include: financial management, personnel management, operations (production) management, procurement and supply, technical-technological aspects of management, etc. Therefore, project Manager needs to have a sufficient level of competence and developed emotional intellect. The challenges facing leaders require the ability to carry out an active search for the necessary decision making information; to recognize the emotional state of the interlocutor, to interpret adequately the content of the received information taking into account their nature, degree of completeness and accuracy, the presence of "hidden meaning", manipulating, etc. The project Manager must be able to interact with their partners, subordinates to achieve the goals, to obtain the desired effect (change of behavior, thoughts, relationships, etc.). How to learn to recognize someone else's experiences? And why empathy is the key to success? This is discussed in the paper.

The purpose is to track how empathy, emotional intelligence and transformational leadership affect the success of a project.

The contribution of this article is a constructed mathematical model of the relationship and the role of the components that affect the success of the project.

The ability to empathy is a professionally necessary quality for all professionals whose work is related to people (TOP managers, line managers, officials, salespeople, personnel managers, etc.). In a business environment, empathy helps to best build relationships between employees and this not only forms the general culture of communication in a company, but also helps to solve business problems more effectively. Transformational leadership makes subordinate leaders. Employees are given relative freedom so that they can independently control their activities within delineated boundaries. They are involved in the problem-solving process and learn new ways of working, which contributes to increased productivity.

Keywords: Simulation · Emotional intelligence · Empathy ·
Transformational leadership · Project success · Management of relations

1 Introduction

The ability to recognize the feelings of another person, and the correct response to her emotions is necessary in very many areas, from trade to social activity.

© Springer Nature Switzerland AG 2020
A. Palagin et al. (Eds.): MODS 2019, AISC 1019, pp. 209–222, 2020.
https://doi.org/10.1007/978-3-030-25741-5_21

To create a product or service that you are interested in, you need to know well who will use them. The general characteristics of the target audience: gender, age and profession are undoubtedly important, but for the best companies in the market, this information about the client is clearly not enough. This requires empathy - the ability to empathize with another person, the ability to put himself in his place.

A manager's job requires a lot of emotional feedback. Emotional managers have a great deal advantage over non emotional ones, therefore that only they are allowed to lead and create new types of projects. They are like usually popular. But, unfortunately, there is also back side of the medal: often our emotions make themselves known in the most inappropriate moment, for example, when necessary restraint and self-control. Conducted studies have shown: emotional states managers are functionally related to them work. Project management is full of activity various tense situations and various factors associated with the possibility increased emotional response.

The topic of the article "Modeling of empathy, emotional intelligence and transformational leadership to the project success" is pretty important for present days. Persons (project manager, participant of project management team) are the main resources of projects and they are a key factor influencing project success. So, understanding of emotional condition of project management team participants is a direct path to project improvements.

2   The Primary Research Materials

The problem of the development of emotional intelligence, paid attention in his scientific studies H. Waisbach and W. Dance, the problem of creating a model of emotional intelligence S.L. Rubinstein, J. Meyer, P. Salovey, D. Caruso, D. Goleman. The organization of the diagnosis of emotional intelligence is considered in the works of K. Jung, R. Cooper and A. Savaf.

The phenomenon of empathy remains the subject of heated discussions by philosophers, ethics, artists, psychologists, educators, ethnologists and doctors. The urgency of the problem of studying empathy is confirmed by the data of empirical studies of modern teachers and psychologists - L.P. Zhuravleva, V.A. Chirichok, O.M. Parkhomenko, O.P. Sannikova. Different aspects of the development of empathy of the individual were covered in the writings of well-known domestic psychologists and teachers (K.O. Abulkhanova-Slavskaya, M.Y. Borishevsky, D.B. Elkonin, O.V. Zaporozhets, G.S. Kostiuk, S.D. Maksimenko, N.I. Nepomyjshch, V.O. Sukhomlinsky and others), which became the methodological and theoretical basis of many studies of this problem. Among them, a significant place is the study of emotional development, sensitivity, promotion and empathy (L.I. Jarnzian, A.D. Kosheleva, L.P. Strelkova). A number of studies were devoted to certain aspects of the development of empathy and its components (L.P. Alekseeva, S.B. Borisenko, L.P. Vygovskaya (Zhuravlyova), T.P. Gavrilova, L.P. Strelkova), the ratio of empathy and morality (A.A. Valantinas, A.V. Solomatina, N.O. Shevchenko).

The substantive content of this phenomenon can be disclosed by appealing to the concepts of "sympathy", "compassion", "empathy" (I. Kant, A. Schopenhauer), which were developed within the framework of philosophy, ethics, aesthetics, psychology

(V. Warringer, V. Dilthey, T. Lipps, T. Ribot, A. Smith, G. Spencer, E. Titchener, G. Scheler). The latest research on empathy issues is O. Bodalev, T.V. Vasilishin, T.P. Gavrilova, T.D. Karyagin, O.M. Parkhomenko, K. Rogers, O.P. Sannikova, M.V. Udovenko and others.

The scientific approach to the study of the phenomenon of leadership has a nearly century history. Did early studies focus on the study of leadership functions such as leadership, control, coordination, and regulation of subactivity, traditionally correlated with the transactional leadership style. Starting in the second half of the 1970s. leadership theory added alongside, but of approaches familiar to the reader in communication, is it with charismatic leader theories (Conger, Kanungo), neo-charisma tactical leadership (Bryman, Sosik, Dworakivsky), transformational (Burns, Bass, Tichy, Devanna) and non-transformational leadership (Yammarino, Bass, Yukl).

3   Emotional Intelligence and Empathy in Project Management

Recently, the term emotional intelligence - the emotional intelligence - is becoming increasingly popular, but in science, there is still no clear definition of this concept.

For the first time, the designation EQ - emotional quotient, the emotional factor, by analogy with IQ - the coefficient of intelligence - was introduced in 1985 by clinical physiologist Ruven Bar-On. Together with Daniel Goleman, the most famous in our country, these scientists form the "three leaders" in research on emotional intelligence. The total number of scientists involved in research in this area is enormous.

The concept of emotional intelligence is probably the only theory in management based on neurophysiology. Emotional intelligence is personality traits that give the ability to recognize and analyze the emotions of surrounding people and their own emotions. Thanks to the developed emotional intelligence, project managers are able to achieve their goals using more flexible behavior strategies [1].

With project managers, different situations occur in life, therefore, it is very important to have the ability to interpret your emotions as negative and positive, in order to use your positive thinking to change your state. That's why it often happens in life that people who graduated from school with a gold medal and have diplomas on higher education do not reach professional heights perfectly, because they have not developed enough emotional intellect. IQ does not provide professional success, it is necessary to have a developed emotional intelligence.

Goleman is available to explain how the almond-shaped body, the emotional center of the brain, affects the activity of the cerebral cortex, which is responsible for logical thinking.

Goleman gives the following information in his book that the most effective project managers are those who can show feelings and mind towards subordinates.

Today it is known that without emotional intelligence, effective leadership is impossible. D. Goleman presents convincing data studies at Harvard University: the success of any activity, only 33% is determined by technical skills, knowledge and intellectual abilities (i.e. IQ), and 67% – emotional competence (EQ). And for heads, these figures differ even more: only 15% of success is determined by IQ, and 85% – EQ [2].

The project manager in interaction with subordinates should take into account their emotional state.

IQ and logic are fully formed before the 17 age, and emotional intelligence develops and improves throughout life.

Modern employers are increasingly inclined to employ those professionals who are quick to be guided in life situations, professional problems, are able to be active, take initiative in their own hands, optimistic about the ability to succeed, courageously and well-balanced approach to planning and implementation of work that can prove the case to logical completion, rebuild (to adapt quickly to changes in the design environment), if required by the right. The most successful in their activities are those people who skillfully combine the mind and emotions [3].

People with high emotional intelligence are:

- make decisions faster;
- operate more effectively in critical situations;
- better manage their subordinates, which, accordingly, promotes their career growth and prosperity of the structure in which they operate.

It is clear that the project manager needs to work with interested parties, subordinates, interacting with them, trying to understand their motives, reasons, or their actions. At the same time, "penetrating into the inner world" of colleagues, it is important not to become the victim of manipulation himself.

Simply purely intuitive is attracted to us by an interlocutor who listens carefully to us, understands our emotions and feelings, does not give any marks to our actions, does not torment us with advice "from our own lives", respects our opinion (even if he does not agree with us). Then there is a strange feeling of unity, a feeling that you are both - at one wave.

It is believed that the concept of "empathy" comes from the Greek pathos - a strong and deep feeling, close to compassion, with the prefix em, which means the direction inside. You can feel empathy, even at the root of disagreeing with the point of view of the interlocutor.

Detecting empathy with your interlocutor means to perceive the inner world of another, but without losing contact with oneself. This means that you must retain the ability to return to your world of emotions. If the shade "as if it is with me" (the key part - "like") disappears, then instead of empathy there is an identification with the emotional state of the interlocutor, you become infected with his emotions and experiences to the same extent as he.

Empathy does not mean "putting yourself at the place of the interlocutor," it's not copying his feelings. Empathy is an attempt to look at things through the eyes of the interlocutor. Another very important point: one can feel empathy, even at the root of not agreeing with the point of view of the interlocutor. That is, you are capable of deep understanding of the feelings of the person with whom you speak, you distinguish your own emotions from those that arose in response to the emotions of the "other side" in the conversation.

Many of you probably have heard of a psychologist Carl Rogers, he defines empathy as follows: "Being in a state of empathy means to perceive the inner world of another accurately, with the preservation of emotional and meaningful shades. As if you become this other, but without loss of feeling "like". Yes, you feel the joy or pain of the other, as he feels, and perceive their causes as he perceives them. But there must be a shade of "like": "as if I'm happy or frustrated."

Empathy is not a formal logic nor an appreciable reaction. Empathy promotes effective communication [4].

Project managers in their work constantly feel the need to understand the interlocutor as fully as possible.

This is felt during an interview with a potential candidate for a vacant position, in solving conflict situations, in the formation of a system of motivation, in the creation of project teams, in optimizing the number of staff, in the release of employees - in all these cases, the project manager is vital to listen and hear interlocutors.

For efficient and effective work with the personnel it is necessary to understand the essence of the real motives of the actions of the employee, the source of his interests, the causes of lies, the goals of isolation.

When forming a team (especially in the "storm" stage), various emotions are raging, which is extremely difficult to control and direct in a constructive way. It is worthwhile making a reservation. It is not about the phenomenon of nature, but about the five stages of team formation: formation, storm, normalization, execution and completion.

If we talk about "storms" in a very short way, then you can characterize it so. Initial optimism after a starting jerk gives way to pessimism, if not frightened by the tasks set. A sense of disappointment or disagreement about the goals, responsibilities in the project.

And in order to achieve a general result, it is necessary not only to listen, but also to understand the point of view of each member of the team, to come to a common opinion, while avoiding a clash of interests and the collapse of the team.

If the company introduces a system of motivation based on an individual approach, then only through an empathic hearing (it is also called "active listening") it is possible to determine the internal motives of each employee, and, therefore, get to the point, making a specific motivational proposal to a unique specialist, expert. And, as a consequence of the competent elaboration of the system of motivation, one can solve the problem of retaining the key specialists of the company. And in our time of high competition, the latter task is especially relevant: as a highly professional specialist, an expert can be a talented leader.

Accordingly, when "good" people go, very often take away all their team.

Business erases state and national boundaries. Now no one will surprise anyone with the multinational staff of the company, as it was 15 years ago. Intercultural features of the company require a project manager special knowledge in the field of relations (and business including) and a deep understanding of the intricacies of the culture of different countries and denominations, whose representatives work in the company. In cross-cultural companies, empathy becomes a connecting link in shaping a common corporate culture.

Microsoft CEO Sathya Nadella argues that empathy is a critical component in developing products or concepts that helps understand the needs of people and gain their trust. In addition, the ability to read alien emotions will be needed in negotiations and for conflict situations. The head of this skill will inspire subordinates and lead them with them. The line-up employee is to maintain a friendly atmosphere in the team (and this, as the research shows, has a positive effect on our performance). By showing empathy in working with clients, you can find a common language even with the hardest people.

Usually emotional exchange occurs on a thin, almost imperceptible level. The ability to calm down the painful experiences of other people and the ability to communicate with the interlocutors, who are in absolute fairy tales, are indicators of higher skill. The only effective strategy is this: you need to deeply absorb the feelings of a person, and then adjust it to a more positive wave.

How to develop empathy? It is important to understand that it is closely related to other elements of emotional intelligence - the ability to recognize and control their own feelings. This should be learned first and foremost.

The map of empathy was created many years ago by Dave Gray, the founder of XPLANE, the author of the techniques of brainstorming "Geyshtorming", the author of books on visual thinking practices.

Empathy map is an idea rendering tool developed by XPLANE that allows you to put yourself at the user's point of view to take a look at the problem your product solves with your eyes. The map of empathy is a scheme in which the center of a representative of a certain user segment is placed; on the different sides of it there are 4 blocks ("think and feel", "say and do", "see", "hear"). The conclusions are presented in two additional blocks: "problems and pain points" and "values and achievements".

The map of empathy is relevant where you need to look at the product from the eyes of the client:

1. Development of strategy;
2. Launching, completing a product or service;
3. Search for new directions;
4. Improving the level of service;
5. Working with the atmosphere in the company [9].

If the project manager is already familiar with the audience, then the empathy map details the context of the use of the product, at the start of the project will show where the gaps in the data. An empathic map can be constructed for any product, and it does not matter if it is implemented, or exists only in the format of the idea.

So, what you need:

1. Determine the approximate target audience.
2. Conduct a "brainstorming" with a team/customer and fill out a map based on your experience and assumptions.
3. Conduct online research. Not only and not so much how users interact with your product, but how they behave in relation to the problem that your product solves. For example, if your product is a project management system for Microsoft Project, then you need to learn what ways the project management tasks are solved by users,

which they like, and what they do not like, which in principle can not be made, and I would like to the tools they involve in the process, etc. But it is important to remember that empathy is different from the impassive study of how a person uses something. Empathy is related to empathy, with the understanding of what the user wants to achieve, no matter if he knows about the product that the team creates.

4. Conduct interviews (potential or real) users. You can also observe their work instead of interviewing (better, in addition to it).

The map of empathy is an instrument of brainstorming. to work in a team was effective, it is important:

- prepare in advance - if they send materials to the participants in 2–3 days, they will have time to get acquainted with the topic, to think about the tasks and to prepare the necessary data;
- speak the main task at the beginning of the discussion - so the project manager will make sure that all participants in one wave know the goals and rules of work.

The information is distributed in blocks as follows (Fig. 1):

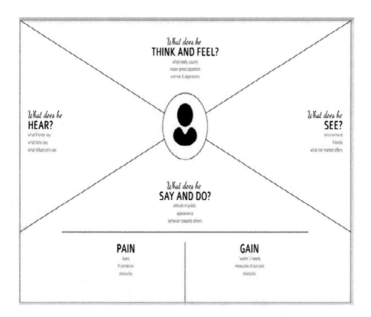

Fig. 1. Empathy card

1. Think and feel: what is there to doubt? It's better to look for this information where users complain: for example, in forums.
2. Say and do: how does the user behave publicly? What does it say? What is the solution to the problem? This information should be searched on social networks.
3. See: what is the environment in which the user is located? What are the suggestions and alternatives your product faces?

4. Hear: how does the environment in which the user is, affects him? What say colleagues, acquaintances, authoritative sources for him? What media channels have an impact on the user? Unlike the block "see", the information here does not necessarily correspond to reality. But the user trusts her. Where to find information for this block: rumors and thoughts on forums.
5. Problems and pain points: what troubles the user? What is he afraid of? What could be the reason that he would give up your product? Often, the "I Think and Feel" block becomes a source of information for this. All these fears and doubts will have to be dispelled, and this can be done by a bunch of ways: from the "right" text in the interface to the individual consultations.
6. Values and achievements: what will help the user to get rid of problems and doubts? For what product features is he willing to pay? What values should be broadcast? Conclusions from this block affect the product from a variety of different parties: they can cause both small changes in the interface or in the text, and the addition/ exclusion of certain functionalities, and sometimes even a change in product positioning.

The main risk of the project is becoming "burnt down" employees. The project rhythm and the large flow of tasks cause such people to be annoyed. To understand why such people appear in an organization, consider the model of development of a specialist, built on the parameters "skill/motivation". This model is needed to understand at what stage the "life cycle" can be staffed by the project, and how it affects their motivation.

Consider the modeling of the personal change. For that purpose using curve of the manager performance during the life cycle of the project (Fig. 2).

In proposed model Empathy to Project Manager appear in critical situation as the motivation factor.

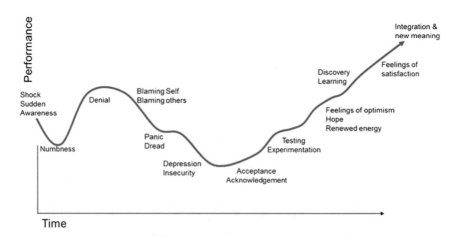

Fig. 2. Curve of personal manager changes during the life cycle of the project

It becomes apparent that at the beginning of the project there is a decline in the team's activity and the manager needs empathy from the stakeholders and a conceptual model must be presented in two parts in order to achieve a balance:

$$Motivation = Emotional\,Intelligence + Empathy \tag{1}$$

$$Project\,success = Emotional\,intelligence + Transformational\,leadership + Competencies\,of\,project\,manager + Values \tag{2}$$

The motivation of the team at the beginning of the project is most affected by the emotional intelligence of employees and the empathy of the stakeholders.

Step 1 - Little experience, lots of enthusiasm (low skill, high motivation). This may be a young specialist who got his first job; a person who has decided to try himself in a new profession or a professional who has been promoted to a managerial position. In general, any of the options when a person just came to a new place is very keen to succeed, but still does not understand how to do it.

Step 2 is the first disappointment (low skill, low motivation). At this stage, our expert comes to the understanding that everything is not as easy as it seemed at first. He makes the first mistakes, succeeds at once, and, most importantly, comes the understanding that the path to the summit is long enough and not at all as simple as it seemed at first.

Step 3 - Natural growth (skills are increasing, motivation is different). If a specialist manages to overcome the previous stage, he enters the path of professional growth. It is already clear what exactly needs to be done for development, it is also clear that the path to success lies through a long methodological work. This stage is usually quite long, with its black and white stripes, so there is not one level of motivation on it, the only thing that can be said is usually enough to keep moving forward.

Step 4 - Competent specialist (high skill, strong motivation). At this stage, the specialist goes to the competence plateau, and can begin to perform tasks autonomously (without a manager), gradually expanding his sphere of responsibility and helping beginners at work. Ideally, having worked for some time in this mode and having prepared a replacement, the employee goes to the raise and returns to the first stage, starting a new turn of his career spiral [10].

In real life, unfortunately, it often happens not so. New posts are not always available, and teaching yourself a worthy replacement does not allow for the absence of any candidates for this post, so the next stage comes.

Step 5 is a very competent specialist. This condition is described by the English word "overqualified" and indicates a significant discrepancy in the competence of the specialist and the needs of his office. The state is characterized by a constant decrease in motivation, due to the lack of a positive connection from the implementation of complex, interesting tasks. After some time, the decline in motivation leads to disappointment from work, and shifting the priority from work to something else (hobbies, families, and outsourced projects). Typical external attributes: a person starts to work strictly on schedule, and at a meeting more speaks not about his work, but about something that took her place in the system of priorities. Man at this stage does not necessarily work badly. On typical tasks, lack of motivation from a good specialist is offset by a high level of professionalism. Problems begin if complex tasks appear that require exit from the comfort zone; in this case there will be behavior that is the opposite of the expected - instead of labor enthusiasm there is a rejection or even resistance. This is because solving problems beyond the competence requires a high place in the system of priorities, and the place is already occupied by something else [5].

Expectations of management begin to disagree with the behavior of the employee.

The final, sixth stage of this process will be professional degradation, and the person from the state "may, but does not want" goes into the state "does not want and can not." Such employees are either cut short by the state's next optimization, or they end their path in "paper" positions with a low level of responsibility.

4   Common Features of Modeling Transformational Leadership and Emotional Intelligence

Recently, the classification of leadership has changed, in particular, Burns includes the following formats:

- Transformation;
- Transactional;
- Non-intervention.

Transformational Leadership. Leaders put the needs of employees higher than their personal.

Transaction leadership has three components:

- Contingent remuneration, while the productivity of subordinates is related to conditional rewards;
- Active management, through which leaders monitor effectiveness and apply corrective actions if deviations occur;
- Passive management (leaders intervene only when problems become serious)

The third type of non-intervention leadership. This leadership style can be described as non-leadership or avoiding leadership responsibility. Leaders do not respond to requests for help, and oppose the expression of their views on important issues.

Leadership is undergoing fundamental transformation today, the transformation from a leader critic to a leader as a partner and coach takes place. This role transformation requires some skills, because leadership is what you do with people, not with them. Numerous studies have shown that transformational leadership positively affects productivity, job satisfaction. Therefore, we can assume that the skills.

The transformational leadership will stimulate efficiency and innovation in this volatile market:

- Idealized influence;
- Inspirational motivation, where the leader inspires and supports team spirit;
- Intelligent stimulation, when a leader encourages creativity and innovation;
- Individual consideration when a leader maintains and supervises every follower.

Transformational leaders use intellectual stimulation to encourage innovative ways of working and problem solving.

Transformational leaders stimulate and inspire followers to achieve extraordinary results, as well as develop their own leadership potential. Transformational leaders take into account the needs of individual followers.

Salovey argued in their first article that there is another type of intelligence called emotional, which can help to better understand who has succeeded than the ratio of mental abilities [6].

Goleman said that 80–90% of success is determined by the presence of emotional competencies [7].

Guided wisely, leaders get incredible value from emotions and the development of real self-efficacy. Emotional intelligence helps leaders make better decisions and receive effective returns from their employees. Scientists have found that cognitive abilities predict less than 2% in efficiency, while 25% of the variation in productivity is explained by emotional intelligence.

The transformational leader demonstrates empathy, motivation, self-consciousness and self-confidence. Goleman described the above subcomponents of emotional intelligence. Emotionally, intellectual leaders use empathy to connect to the emotions of other people they lead. These leaders sympathize with and also express emotions that are group experiences. Thus, the team feels that the leader understands it.

Charisma, a trait of transformation leader, is well-developed social and emotional skills. Emotional intelligence is both a basic and a necessary component of personal charisma demonstrated by transformational leaders. Transformational leaders use emotions to communicate and motivate followers.

Sosik proposed four points in which the emotional intelligence and transformational leadership intersect:

- Compliance with professional standards of conduct and interaction that are associated with idealized influence or charisma;
- Self-motivated, as ability to control;
- Intelligent stimulation: a leader must be able to stimulate professional and intellectual development. Bass has established trust as the main component of the transformational style of leadership. And Cooper offered trust as an important characteristic of emotional intelligence;
- Individual focus on others [8].

The level of emotional intelligence leaders regulates their ability to control the feelings and emotions of teams and motivate them to achieve their goals. Such leaders inspire their teams through positive thoughts and a clear vision.

Each manager has the ability to develop the emotional competence of the team and become a resonant leader. Leaders with high emotional intelligence are self-conscious and they understand themselves, they are sympathetic and attentive. This resonance comes naturally to emotionally intelligent leaders, and this resonance boosts productivity. Caruso argued that the exact recognition of emotions in others is critical to the ability of leaders to inspire and build relationships. Empathy precedes the transformational behavior of leadership. Such leaders also develop team spirit, role modeling enthusiasm, high moral standards, integrity, optimism and meaningful work for followers.

Rather, he studied the relationship between leadership style, intuition and emotional intelligence for women and men leaders. He found that female managers show transformational leadership behaviors more often than men and have higher levels of emotional intelligence and intuition. Intuition correlated significantly with the emotional recognition and expression, and the emotions of direct cognition.

5   Possibilities of Developing Empathy and Application of Results in the Project

As mentioned above, we understand others by realizing their own emotions. If you have already mixed up with your feelings, you can go to the training of empathy. To do this, use the following tools.

Many of us listen indifferently, trying to do something else at the same time or plunge into their own thoughts. An amputation hearing occurs when we completely devote ourselves to another person. Here's how you can develop this skill:

1. Allow others to speak. You do not need to provide verbal or non-verbal tips and to arrange for a person a proposal. When we interfere, we inadvertently direct the interlocutor to what we want to hear, instead of letting him talk in the right direction.
2. Give the speaker an absolute attention. Perhaps you know how it is when a person is doing other things while he is listening to you. Whether he is reading something on a computer or looking over your shoulder, wanting to see someone, this seems like a disgrace.
3. Play and generalize. Periodically repeat what you have heard to make sure your understanding is accurate. It offers the speaker an opportunity to reformulate or explain some thoughts. For example, you can say: "Yes, right, did I understand the situation? Do you feel disregard for the boss when he does not say positive about your contribution to the common result?"
4. Focus on emotions. Hearing with empathy involves interpreting the thoughts and feelings of others. In addition to playing and heard generalization we can also add sympathetic words to what they say, such as "Sounds depressing" or "You look angry."
5. Put yourself in someone's place. Do not interfere with the interlocutor from the position of advantage: "Guy, that's what you have plucked". Instead, think like this: "How is it, to be in place of this person right now? What would I feel in this situation?"
6. Reject judgments. The hearing about empathy involves refusing our own judgments, needs, and priorities and concentrating on another person.

What we do with the information received during the hearing is also important. Often, there is a temptation to recall its own story and suggest a solution to the problem. This method is not very empathic. A more effective approach is simply to say: "It seems difficult," "How can I help?" Or "What support do you need right now?"

6   Conclusions

EQ is a necessary factor in enhancing mental skill, because the recognition of their feelings and manage them in a constructive way increases the intellectual power of the individual. To increase the level of emotional intelligence may be possible, but not through traditional training programs aimed at the part of the brain which controls our rational ideas, and long practice, feedback from colleagues, and personal enthusiasm in the desire to change yourself is an essential step to improving EQ, and as a result, successful self-realization.

Actively using empathy in the exchange of information, we are trying to appropriately adjust the receiving side and adapt the option of coding and transmitting the message to the individual or group and the situation.

A successful application of empathy can significantly reduce the possibility of misunderstanding when decoding a message by the receiving person.

How to learn empathy? Start with yourself. Live in harmony with yourself, be positive and transfer your mental comfort and positive feelings and attitudes toward life in relationships with friends and colleagues around you, look for individual ways of interaction that will help to unleash the potential (both yours and your interlocutor), to find resources for fruitful cooperation.

Sometimes people tend to confuse pity and empathy. And if pity can damage business, then empathy is not. Empathy does not imply indifference. On the contrary, it is an understanding of the complicated, tragic (or, conversely, happy) situation of the employee and the ability - if necessary - to help him.

Transformational leadership enhances the motivation, morale and productivity of followers through various mechanisms. In this model, the leader is a role model for his followers.

As a result, this article describes how emotional intelligence, empathy, and transformational leadership influence the success of a project.

References

1. Jakobony, M.: Reflected in people. Why do we understand each other?. United Press (2011)
2. Goleman, D.: Emotional intelligence in business - Mann, Ivanov and Ferber, Moscow, p. 356 (2013)
3. Rusan, N., Bushuyev, S., Bushuyev, D.: Emotional intelligence—the driver of development of breakthrough competences of the project. In: International Scientific and Technical Conference on Computer Sciences and Information Technologies (CSIT), pp. 1–7. IEEE, Lviv, 5–8 Sept 2017
4. Sargwelladze, N.: On the Balance of Projection and Introjection in the Process of Empathic Interaction // Unconscious: Nature, Functions, Methods of Investigation: 4 t. / By common. edit Prangishvili, A.S., Sharoziya, A.E., Bassin, F.V. T.3 Tbilisi, pp. 485–490 (2009)
5. Decety, J., Ickes, W.: The Social Neuroscience of Empathy (Social Neuroscience). MIT Press, Cambridge (2011)
6. Brackett, M., Rivers, S., Salovey, P.: Emotional intelligence: implications for personal, social, academic, and workplace success. Soc. Pers. Psychol. Compass 5(1), 88–103 (2011)

7. Goleman, D.: Emotional intelligence at work lane. From English. A.P. Isaeva. – M.: AST, Moscow. , p. 476 (2010). Vladimir:VKT
8. Sosik, J.: Full Range Leadership Development, 1st edn, p. 424. Routledge, New York (2009)
9. James, M.B.: Transforming Leadership, Reprint edition, p. 240. Grove Press, 30 January 2004
10. Mayer, J.D., Salovey, P., Caruso, D.R.: Emotional intelligence: new ability or eclectic traits? Am. Psychol. **63**(6), 503–517 (2008)

Management of Online Learning Through Modeling of Non-force Interaction of Information Agents

Nataliia Yehorchenkova[1]([✉])(ⅈD), Iurii Teslia[2](ⅈD),
Oleksii Yehorchenkov[2](ⅈD), Liubov Kubiavka[2](ⅈD),
and Iulia Khlevna[2](ⅈD)

[1] Kyiv National University of Construction and Architecture,
Povitroflotsky Avenue 31, Kiev 03037, Ukraine
realnata@ukr.net
[2] Taras Shevchenko National University of Kyiv,
Volodymyrska Street, 64/13, Kiev 01601, Ukraine

Abstract. The purpose of the article is to demonstrate the possibilities of applying the non-force interaction theory for developing management systems for online learning. The analysis of problems of online learning and ways of their solution is conducted. The main approaches to the implementation of online learning are considered. It is shown that in order to increase the effectiveness of online learning, it is necessary to set up learning programs for the knowledge of each trainee. This is possible only through the use of modern online learning management systems that could take on the functions of developing individual programs for each trainee. It is proposed for the first time to use the non-force interaction theory to create such online learning management systems. The conceptual and mathematical apparatus of the non-force interaction theory in the application to the learning processes is expounded. It is shown that the use of tools of the non-force interaction theory in online learning will improve the efficiency and quality of the educational process through managing the positive attitude of the trainee. The interacting subjects of the learning process are singled out and a model for changing the attitude to reality in the course of non-force interaction is developed. It is shown that on the basis of a non-force interaction model, it is possible to create fundamentally new intelligent systems capable to form an individual trajectory of student online learning.

Keywords: Online learning · Non-force interaction · Artificial intelligence

1 Introduction

Online learning appeared in 60th years, when the University of Illinois initiated a classroom system based in linked computer terminals where students could access informational resources on a particular course while listening to the lectures that were recorded via some form of remotely linked device like a television or audio device [1] Today, among scientists there are differences in the definition of «online learning» .

Authors [2] describes the differences and presents main views on «online learning» :

© Springer Nature Switzerland AG 2020
A. Palagin et al. (Eds.): MODS 2019, AISC 1019, pp. 223–233, 2020.
https://doi.org/10.1007/978-3-030-25741-5_22

Some prefer to distinguish the variance by describing online learning as "wholly" online learning [3], whereas others simply reference the technology medium or context with which it is used [4]. Others display direct relationships between previously described modes and online learning by stating that one uses the technology used in the other [5, 6]. Online learning is described by most authors as access to learning experiences via the use of some technology [7–9]. [7] and [9] identify online learning as a more recent version of distance learning which improves access to educational opportunities for learners described as both nontraditional and disenfranchised. Other authors discuss not only the accessibility of online learning but also its connectivity, flexibility and ability to promote varied interactions [3, 10, 11].

We will use definition of authors [12, 13] who identify online learning as a form of distance learning or distance education, which has long been a part of the American education system, and it has become the largest sector of distance learning in recent years.

Online learning has big impact to American society. In a paper [14] highlights, that for last decade online learning is one of the fastest growing trends in educational uses of technology. By the 2006–2007 academic year, 61% of US higher education institutions offered online courses [15]. In fall 2008, over 4.6 million students—over one quarter of all U.S. higher education students—were taking at least one online course [16]. In the corporate world, according to a report by the American Society for Training and Development, about 33% of training was delivered electronically in 2007, nearly triple the rate in 2000 [17].

Due to the increasing popularity online learning and its ever greater penetration into the sphere of education of the whole world appear a strong interest from the scientists side in questions of measurement, collection, analysis and reporting of data about students with aim of understanding and optimization of learning and conditions in which it occurs.

2   Methodological Basis of Personalized Learning

Typically, the system of knowledge acquiring is realized in the presence of learning subject and object. Traditionally, an instructor is a subject and a trainee is an object. In the classical approach learning process is a purposeful interaction of instructor and trainee in which the trainee receives knowledge. But only the existence of such a pair does not ensure their cooperation and does not lead to a functioning of a system "trainee – knowledge". This system takes effect only when a certain attitude towards obtaining knowledge is arisen in a trainee. So, it is important to analyze the traditional learning models to distinguish their advantages and disadvantages in terms of information influence and then form requirements for a training model that will personify the benefits of existing models and form a non-force influence on trainee for the system "trainee – knowledge" operation.

Traditional planning and organization of educational process is based on the curriculum, which included the separated schedule of the educational process; the discipline and the number of hours allocated to lectures, laboratory and practical seminars, independent work of students; deadlines for assignments; preparation of tests and examinations; examination session; exams passing, preparing diploma, vacation. Accordingly, information influence on the trainee is carried out in accordance with the educational plans and work programs. This educational model adopted in most schools in the world and operates under the slogan "We teach – therefore he knows". This model has both positive as negative information influence on trainee. Its features are presented in Table 1.

Table 1. Traditional learning model features ("We teach – therefore he knows")

Processes	How processes are implemented	Process estimation	
		Positive	Negative
Academic learning	The instructor explains the material, regardless of whether the trainee knows the material or not	If the information was unknown the trainee receives information and reveals initial understanding	If the information was known, disinterest to the course is formed
Practical training	The instructor gives practical, laboratory work and seminars	If the information was unknown the trainee interprets, deepens training information	If the information was known the trainee has critical attitude about the time spent on learning information
Educational process preparation	The instructor prepares the same course document package for all trainees	Save instructor's time for preparation for classes	The individual needs and characteristics of each trainee is not taken into account
Evaluation	The instructor evaluates the trainee's knowledge	Default and an objective in most cases form of control	End-knowledge of the student, not the quality of efforts to obtain new knowledge is taken into account

Taking into account these results (Table 1), it can be argued that the system "trainee – knowledge": "We teach – therefore he knows" is not quite rational and functional because the trainee's interests as a subject are not taken into account and ability of the trainee to acquire knowledge, skills, experience knowingly and intentionally is not expected. The trainee as an individual is limited in determining his possibilities and maximalization of self-identity. The trainee physiological characteristics is not considered. Of course, independent work of students helps in this direction. But, it is standardized and the same for everyone in this model.

It is advisable to create a such model of study that will combine the existing model (mixed, open and distance) and implements them on the basis of scientific predict of the consequences of the information influence on learning objects. And that could be adapted to the capabilities of each trainee. The basis of this model is a description of information influences that will be used to determine boundaries of the trainee knowledge with further influence to him using incentives to overcome non-acquaintance and the formation of individual education plans for each trainee with the creation of a positive attitude from their learning. That is, this model should provide personal approach to each trainee.

Individual training is a model of educational process organization in which the instructor interacts with only one student, or one student interacts only with the means of learning, or two students interact with each other (reciprocal learning) without teacher [18].

Definition 1. Individualized learning model is a model that represents the existing learning models and provides identifying individual needs of the trainee through collaboration instructor and trainee. This learning model operates under the slogan "He does not know, so we teach". Positive and negative aspects of this model are shown in Table 2.

Table 2. Individualized learning model features ("He does not know, so we teach")

Processes	How processes are implemented	Process estimation	
		Positive	Negative
Academic learning	The trainee is tested. The test defines the trainee knowledge level. The instructor with individualized plan informs of new, not known for trainee knowledge, explains	The trainee accepts only new information, reveals initial understanding	Significant amount of work for the instructor
Practical training	The instructor gives practical, laboratory work	Knowledge and skills in the context of new tasks for trainee are formed	Significant amount of work for the instructor
Educational process preparation	The instructor prepares personal work programs	Individual needs and characteristics of each trainee are taken into account	A large amount of information the instructor needs to prepare
Evaluation	The instructor evaluates the trainee's knowledge	Not only finite knowledge, but the quality of student efforts is evaluated	Different educational material is evaluated so lost meaning in comparing the obtained evaluations

An individualized model of learning allows to increase the probability of memorizing the material by a person. Consider this with an example. The general population $n = 30$ trainees, which is characterized by homogeneity and stability by the IQ Test, is randomly divided into three equal group of 10 ($n_1 = n_2 = n_3 = 10$) trainees. It was offered to the first sample ($n_1 = 10$) to solve the problem in some professional area.

According to the results of the performed tasks, it was found that only 2 respondents fully performed the task. The second sample (n_2 = 10), before the experiment was trained in the traditional method and the system "trainee – knowledge": "We teach – therefore he knows". After training, the sample was asked to perform the same task as the first sample (n_1 = 10). Found that 6 out of 10 trainees completed the task. With the third sample (n_3 = 10), the experiment was followed by an individualized method and the system "trainee-knowledge": "He does not know, so we teach". After training, the sample was asked to perform the same tasks as the first and second samples (n_1 = 10, n_2 = 10). It is established that 9 trainees fully performed the task.

The study shown that knowledge has changed the probability of tasks performing by the person. If the enterprise randomly selects a trainee from a non-trained sample, his probability of solving a professional task would be 2/10 = 0,20. But with the training, this probability will be equal to 6/10 = 0.6 (for traditional training) and 9/10 = 0.9 (for individualized training). Using an individualized learning model in this example allows to increase the probability of solving problems in a professional area from 0,2 to 0,9.

On the same principle, a study was conducted with one randomly selected student from a group of students who began to study a new discipline. The student is offered to solve 100 professional tasks at the beginning of the course. He solved only 20. After completing the course, according to an individualized model of training, he received the same tasks, with a proposal to solve them. And his result is 90 solved tasks. The knowledge that he received has changed the probability of the correct solving of professional tasks from 0,20 to 0,90. Accordingly, the probability of functioning of the "trainee-knowledge" system has also increased.

Also, the experiment revealed that before the beginning of a new course, it is necessary to take into account the characteristics of each student. The new course must begin with the eliciting contradiction between the knowledge and unknowledge of the trainee, the identification of the trainee weaknesses, the awareness of the psychological, personal needs of the trainee, the definition of methods of perceiving information of the trainee, the level of memorization of the material.

As can be seen from the following examples, learning is an influence that changes the probability of the correct solution of professional tasks. And these questions are very well solved in the non-forced interaction theory. Therefore, the use of this theory tools will be rational. These tools will ensure the functioning of the individual learning technology and will be a driving force behind the formation of a new model of education. But, as part of the traditional organization of the educational process, such training is impossible, because of considerable work costs for the formation of individualized educational programs. It is impossible to do this without the use of special methods and techniques of automation. So, it is necessary to create a computerized system that will assume all the functions of automatically compiling of individual training plans. Therefore, for the operation of the model of individualized learning it is necessary: develop an information system to manage trainee knowledges (ISMTN), integrate ISMTN into an online learning platform, develop a model for a friendly instructor and trainee collaboration, develop a model for distance learning between instructor and trainee, develop a model of trainee's motivation for self-development and self-education.

Thus, an individualized model of learning is a complex concept. Such a model combines pedagogical, informational and managerial components for the formation of a professional in a certain area of activity for a minimum of expenses and a maximum benefit. Therefore, for the implementation of such a model of training, it is necessary to formulate a specific methodology for managing the model of individualized training, to select the appropriate tools and to adjust them to the conditions of the educational institution. After all, the volume of work on an individualized model of training is so significant that it is necessary to implement part of the functions using the tools of modern information systems and technologies in the context of online learning. It is necessary to create a platform that will include not only learning tools but also online learning management tools. Such a managed platform will help create the conditions for the development of the individual needs of the trainee for the possibility of presenting information in the form of video, audio, text, the possibility of contact with the instructor, the possibility of automation of processes and their management.

The formation of a platform for online learning should embody simplicity, convenience, versatility and the ability to integrate with the ISMTN. Creating an effective trainee's knowledge management infrastructure is one of the main principles of an individualized learning model. It is about a system that can form such an individualized learning program. The basis of such a system will be the calculation of the influence on the trainee, so he would receive the required level of knowledge for obtaining competent and professional skills in solving practical problems.

For this purpose, it is proposed to develop a scientific and methodological basis for the use of influences management tools to improve the effectiveness of the online learning process. At the basis of this is the non-forced interaction theory [19]. Let's consider this question in more detail.

3   Application of the Non-force Interaction Theory for Creation of the Comfortable Learning Environment

The non-force interaction theory was widely used in different areas of management, including team management. A conceptual methodological model of non-force interaction of elements of the organization's cultural space in the form of stakeholders was developed, as an object of project management to support the organization's development processes. Models of non-force interaction of elements of the environment of activity in the absence and availability of information about their behavior are built. A model of state mismatch is constructed for different manifestations of interaction counterparties. Also in the non-force interaction theory the questions connected to the project management organization are probed.

But so far, researchers have avoided the issue of managing online learning through modeling non-force interaction between instructors and trainees. Although all previous studies have revealed significant potential of the theory for the construction of specific models in various fields of human activity.

Traditional development methods of online learning materials are, as a rule, expensive, take away a lot of time and require specialized skills which are often difficult for acquiring. For receiving advantages from online training authors recommend to realize this idea through simulation of non-force interaction instructors on trainees, and creations on this basis special tools of online training management.

For the solution of this problem on the basis of the non-force interaction theory it is necessary to follow these steps: (1) to carry out stakeholder analysis (instructors and trainees) from line standpoint of friendliness or antagonism of the attitude to the subject of study; (2) to calculate interdependency of communications, influences and changes in interests of trainees; (3) to manage the non-force influence on trainees.

Traditionally, education is aimed at the formation of knowledge and skills of people who, having become competent specialists, develop their country. The theory of non-force interaction in education declares a value approach. Education does not just give knowledge and skills, it makes more likely the right solution of tasks in the field of professional activity, and hence obtaining a positive attitude in the life of an educated person. Indeed, we want positive life to be more than negative.

Now we will consider training through the non-force influence on trainees which, in to perspective, makes them more or less happy (creates in the long term more or less positive and negative in life). The education system itself provides that the knowledge gained by students will provide an opportunity for them to work better, get higher wages, etc., and, consequently, get more positive in life.

Supersaturation of the country by economists, financiers, lawyers, etc. leads to unemployment and eventually to the performance of work not by specialty. This means that time and money for training are lost, and nothing is received in return, which ultimately makes such people less happy in life. This also leads to poor-quality learning, or learning "theory" without regard to practice.

Very few people from trainees ponder upon it. They are guided by the fact that in the future during of training they will receive many times positive attitude. The positive is necessary to trainees now, during training and as it is possible in bigger volume.

All of us are a part of the educational sphere. We are Teachers. We a single whole with whom we teach. Also, we understand that receiving a positive in activity is possible through optimum communication corporate and personal - when achievement of each of us provides development of education, and development of education is a source of satisfaction of vital needs of each of us that in turn, develops the country and creates positive attitudes for many-many people.

How to make that training activity was positive? Something was already told about it in the subsection devoted to development of methodology of the non-force training. For example, the training material shall be new and interesting. Means, it is necessary to teach not all equally, but all individually.

If there are a positive and a negative from on-line training, then reflexes can be developed. All stakeholders of online training want to receive more positive. For example, if actions in some direction are brought by a negative, and in opposite – a positive, then their relation to training shall change so that the following actions brought more positive. Each stakeholder of online training shall learn to arrive so that to receive a maximum of a positive and a minimum of a negative. At it the reflexes carrying to a positive shall be developed. The model of interaction of stakeholders of online learning is presented very harmoniously. From the inability and ignorance of how to act at the moment when the project starts, through constant learning and development, to experiencing more positive things in the learning process. Therefore, the only criterion for determining the "correctness" of such a movement should be the learner's attitude from interacting with the teacher.

If this method "works" also in the educational environment, then it will become one of arguments of acceptance of universality of the non-force interaction theory.

For implementation of such approach (through a positive in attitudes to results of training) it is necessary to develop conceptual model of interaction and influence of two and more stakeholders of online learning in case of implementation of the given steps of the non-force impact on quality of training.

Originally all stakeholders are defined. In one way or another all of them interact among themselves: communicate, perform tasks, etc. Thereof at each of them positive and negative attitudes from these interactions are created (preparation of a training material, interaction with the trainee, a study of a training material, etc.). And at each of stakeholders the positive or negative manifestations can prevail, and also can be neutral. They are created due to readiness to put resources (time, awareness, etc.) in training and confidence in its successful completion, achievement of necessary quality, obtaining desirable effect. From the point of view of effective management of simulation of the non-force interaction of stakeholders, it is necessary to work out at them "drift in the direction of training". Transition to such status is carried out at the third stage due to implementation of the non-force influences in relation to each trainee (individually). As a result, quality of educational process through control of positive attitude of the trainee can adjust if necessary, these influences and use them as one more additional instrument of control of online training. But it is possible to realize it only in that case when the productive program and information tools capable to undertake functions of determination of an optimum path of movement to the trainee on a training material and its implementation is developed.

The importance of the non-force interaction theory is not only in giving a formal mathematical apparatus for the processes of information interaction management, but also in underlying the creation of reflex intelligent systems. Such systems act by analogy with the human brain. They learn, analyze, classify, react and on this basis, they solve many practical problems. At the basis of such systems is the formation of reflexes for non-force (information) influence.

The authors suggest a method for constructing a reflex expert system for managing individual online training. Such a system will reflect the knowledge of trainees in some software and information environment, which provides the forming of necessary reactions. Let us introduce several definitions

Definition 2. Informed agent (IA), the model of the trainee in the information base of the reflex expert system of individual online learning management (RESL) which is capable to form the RESL reaction.

In fact, the informed agent is an interrelated collection of data and knowledge reflecting the real or abstract trainee, as a result of the interaction of elements of which the RESL reaction is formed.

Definition 3. RESL reaction is a course unit (topic, lecture) that must be studied by the informed agent.

RESL reactions can be set by the teacher, and can be formed independently by the informed agent. It is necessary to create a RESL, consisting of informed agents, each of which defines a certain learning scenario. As soon as a new information enters the system - the results of testing or interviewing/questioning, a new model of the learner, a new informed agent should be formed. The new informed agent generates such reactions, which are determined by the intersection of this informed agent, with other informed agents of RESL for whom the reactions are identified in the learning process.

The reaction is the course unit that must be studied by the informed agent. Reaction can be formed by RESL in the process of interaction with an expert (teacher). And also, can be formed by RESL on the basis of experience about reactions that were formed for other IAs.

Definition 4. Under the **management of online learning**, we will understand the automatic definition of RESL course units (topics, lectures) that need to be studied by the trainee in the course.

For each informed agent, several reactions can be formed. The order of implementation of the course units that need to be studied is determined by the work program of the course. Consider this for an example. Let there be 5 topics in the work program: r_1, r_2, r_3, r_4, r_5. RESL's reactions, which are formed by the information agent are topics r_1, r_3. In this case, r_1 topic will be studied at first and then the topic r_3.

Because the order in the work program is exactly from the first topic to the fifth. There are two stages in the technology the RESL operation.

1. Stage of training

The trainee passes the test (maybe he is also interviewed). The instructor determines which topics should be learned and in which order. What you need to learn is the reaction of an informed agent. Statistics are accumulating: what are the test results (interviews) - what needs to be learned. In essence, the system becomes expert (reflex expert systems for managing individual online learning), as the teacher lays out his understanding of the relationship between test results and questionnaires/interviews and the form and content of training. After some time, learning can be repeated. Consider this technology by procedures.

1.1. Information agents test results and/or interview results are put into the RESL.

1.2. For each information agent, the instructor indicates which course unit he should study (the reactions of the RESL are indicated).

2. Stage of management

Using this technology, it is possible to create systems that form individual trajectories of online learning of students based on the analysis of interview and test results that need to be taken at the beginning of the class, and then in the process of studying the course.

4   Conclusions and Prospects for Further Research

The paper shows that on basis of the non-force interaction model it is possible to create fundamentally new intelligent systems capable of forming an personal trajectory of online student learning. The advantage of such a system is the simplicity and transparency of the algorithm. There is no need for intellectual analysis of data with the involvement of experts, for creation cumbersome databases, for using linguistics.

The online learning management model, demonstrated in this article, is a clear example of the practical applications of the non-force interaction theory. And we hope that this article will help many researchers apply the knowledge of authors for further progress in automating the process of online student learning.

References

1. Woolley, D.R.: PLATO: The Emergence of Online Community (2013). Thinkofit.com
2. Joi, L.: Moore, Camille Dickson-Deane, Krista Galyen e-Learning, online learning, and distance learning environments: are they the same? Internet High. Educ. **14**, 129–135 (2011)
3. Oblinger, D.G., Oblinger, J.L.: Educating the net generation. EDUCAUSE (2005). http://net. educause.edu/ir/library/pdf/pub7101.pdf
4. Lowenthal, P., Wilson, B.G., Parrish, P.: Context matters: a description and typology of the online learning landscape. In: AECT International Convention, Louisville, KY. Presented at the 2009 AECT International Convention, Louisville, KY (2009)
5. Rekkedal, T., Qvist-Eriksen, S., Keegan, D., Súilleabháin, G.Ó., Coughlan, R., Fritsch, H., et al.: Internet Based e-learning, Pedagogy and Support Systems. NKI Distance Education, Norway (2009)
6. Volery, T., Lord, D.: Critical success factors in online education. Int. J. Educ. Manage. **14**(5), 216–223 (2000)
7. Benson, A.: Using online learning to meet workforce demand: a case study of stakeholder influence. Q. Rev. Distance Educ. **3**(4), 443–452 (2002)
8. Carliner, S.: An Overview of Online Learning, 2nd edn. Human Resource Development Press, Armherst (2004)
9. Conrad, D.: Deep in the hearts of learners: insights into the nature of online community. J. Distance Educ. **17**(1), 1–19 (2002)
10. Ally, M.: Foundations of educational theory for online learning. In: Terry (ed.) The Theory and Practice of Online Learning, 2nd edn., pp. 3 – 31. Athabasca University, Athabasca (2004). http://desarrollo.uces.edu.ar:8180/dspace/bitstream/123456789/586/1/Theory%20and%20Practice%20of%20online%20learning.pdf#page=227
11. Hiltz, S.R., Turoff, M.: Education goes digital: the evolution of online learning and the revolution in higher education. Commun. ACM **48**(10), 59–64 (2005). https://doi.org/10.1145/1089107.1089139
12. Bartley, S.J., Golek, J.H.: Evaluating the cost effectiveness of online and face-to-face instruction. Educ. Technol. Soc. **7**(4), 167–175 (2004)
13. Evans, J., Haase, I.: Online business education in the twenty first century: an analysis of potential target markets. Internet Res. **11**(3), 246–260 (2001). http://doi.org/10.1108/10662240110396432

14. Means, B., Toyama, Y., Murphy, R., Baki, M.: The Effectiveness of Online and Blended Learning: A Meta-Analysis of the Empirical Literature (2013). https://www.sri.com/sites/default/files/publications/effectiveness_of_online_and_blended_learning.pdf
15. Parsad, B., Lewis, L.: Distance Education at Degree-Granting Postsecondary Institutions: 2006-07. National Center for Education Statistics, U.S. Department of Education, Washington, DC (2008)
16. Allen, I.E., Seaman, J.: Learning on demand: Online education in the United States (2009). http://www.sloanc.org/publications/survey/pdf/learningon demand.pdf
17. Paradise, A.: 2007 State of the Industry Report. American Society of Training and Development, Alexandria, VA (2008)
18. Onishkiv, Z.: Personalization of the learning process as a scientific and pedagogical problem. Scientific notes Ternopil State Pedagogical University. Series: Pedagogy. No. 9, p. 6–9 (2002)
19. Teslia, I., Popovych, N., Pylypenko, V., Chornyi, O.: The Non-force interaction theory for reflex system creation with application to TV voice control. In: Proceedings of the 6th International Conference on Agents and Artificial Intelligence, pp. 288–296 (2014). https://doi.org/10.5220/0004754702880296

Projects Portfolio Optimization for the Planned Period with Aftereffect Consideration

Igor Kononenko$^{(\boxtimes)}$ ⓘ, Anhelina Korchakova, and Pavel Smolin ⓘ

National Technical University "Kharkiv Polytechnic Institute",
2 Kyrpychov Street, Kharkiv 61002, Ukraine
igorvkononenko@gmail.com

Abstract. The purpose of the article is to create the project portfolio optimization mathematical model for the planned period, which would take into account the receipt and expenditure of funds over time, the social effect of the portfolio, the risks associated with its implementation, restrictions on the availability of funding, and the necessary sequence of projects realization. The existing papers did not solve this problem. We consider a portfolio of projects, each of which can be started at an arbitrary time during the planned period. In this task, it is proposed to take the number of objective functions into account. The first objective function reflects the difference between revenues and expenditures in the implementation of the projects portfolio for projects launched during the planned period. The following objective functions assess the risks and the social effect of the projects portfolio implementation. The task model contains a number of limitations. Available funds must cover the needs of the portfolio in each period. The second constraint of the model requires every project to start no more than once. The third constraint specifies the order of interrelated projects fulfillment. The last constraint allows specifying the requirement of the mandatory inclusion in the portfolio of some project on the time segment. We propose the mathematical model of the project portfolio optimization for the planned period, which takes into account the aftereffect of previously made decisions. The proposed problem pertains to the multicriteria dynamic Boolean programming problems.

Keywords: Project portfolio · Model · Optimization · Planned period · Aftereffect

1 Introduction

Companies engaged in the development of information technology do not fulfill individual projects, as a rule, but a number of projects in a complex sequence. The task of projects selection and an optimal projects portfolio formation for a certain planned period is very important. Not only the individual projects and the whole portfolio success depends on its solution, but also the company's development in the medium and long term does. A large number of goals pursued by stakeholders and managers overcomplicate this task solution, as these goals often contradict. For instance, the

© Springer Nature Switzerland AG 2020
A. Palagin et al. (Eds.): MODS 2019, AISC 1019, pp. 234–242, 2020.
https://doi.org/10.1007/978-3-030-25741-5_23

desire to get more of the income, profit, to ensure self-sufficiency can come into conflict with the organization's social goals. In addition to the goals that are to be optimized, many restrictions under which one has to implement a projects portfolio should be taken into consideration.

Projects in a portfolio can be both independent of each other or in a complex way depend on previously fulfilled projects. The next project can use the results and products of the previous one. The experience gained by the team in the realization of certain projects may allow starting a more complex project that the team could not be able to execute before. Projects portfolio optimization must consider these dependencies.

The projects portfolio can be carried out both at the expense of the organization and at the expense of funds earned in projects. Moreover, the organization at certain stages may need funds earned in the portfolio. Often, funds earned in one project go to cover expenses in another. These relationships should also be taken into account in the portfolio optimization model.

In any case, in each period there should be enough funds for the projects portfolio realization.

A solution of portfolio optimization problem originates from workings in the field of securities optimal portfolios formation.

In optimizing portfolios of securities, as a rule, a two-criterion problem with constraints is considered. The first objective function reflects the income from the portfolio, the second – the risk associated with its implementation. The risk is estimated using standard deviations for income and fund correlations. The problem considers various restrictions, which comprise a limit on financing, including a top and bottom limit, a limit on the funds quantity, a limit on investing in a certain class of assets, a restriction requiring the investment to be a multiple of a certain minimum value, a limit on operating expenses, a limit on the between-assets proportions [1].

The work [2] gives the classification of the most significant papers in the projects portfolio formation field. It shows that mathematical programming methods used to solve it contained the linear, nonlinear, integer, dynamic, stochastic, and fuzzy programming.

The paper [3] considers the mathematical model of the projects portfolio optimization problem, which applies the Boolean variables x_{ij} equal 1, if the i-th project starts in a period j, otherwise, equal 0, $i = 1,...,N$, $j = 1,...,T$. The task's objective function takes the form of

$$\text{Max}\,Z = \sum_{i=1}^{N} \sum_{j=1}^{T} a_i x_{ij},$$

where a_i stands for the potential profit from the i-th project.

Among the limitations that may be considered, the following were taken into account:

the project cannot be started twice,
costs should not exceed the allowable costs for a given period,
other resources must also not exceed the allowable resources in certain periods of time,
all selected projects must be completed within the planned period,

the certain sequence of fulfillment can be set for projects,
some projects may be mandatory and must be included in the portfolio,
only one of the mutually exclusive projects can be selected.

The [4] proposes the projects portfolio optimization static problem mathematical model, which states that all selected projects must start in one year. However, the benefits of each project are calculated by the year of its finalization. The model feature is that the effect function of interdependent projects is taken into account if they are realized jointly, aiming to maximize the total benefits of the portfolio. The model contains a restriction on the number of certain category projects, which are included in the portfolio. Another characteristic of the model is the presence of restrictions on the certain type of resources consumption and the benefits from the projects by the years of the planned period. Both the resources and benefits in the constraints are discounted over time. The task appertains to the tasks of Boolean linear programming.

The [5] lists the objectives and restrictions, which are usually applied to projects portfolio optimization. These include the portfolio overall strategic value and its potential profitability maximization, synergistic effects from projects, available resources limitations, logical connections between projects, the balance of projects categories, risks, and payback periods. It emphasizes the importance of determining the start time of projects.

In [6] it is indicated that when optimizing project portfolios, the net present value (NPV) is considered among the objective functions, with restrictions and dependencies for resources, technical and market interactions are taken into account. Authors note that most of the works assume the simultaneous start of projects in the portfolio. In order to allocate scarce resources, it is necessary to optimize the portfolio accounting for the possibility of setting individual project start times.

The [7] proposes the projects portfolio optimization two-criteria task model. The first objective function reflects the total benefit of the projects portfolio implementation. Experts evaluate the profitability of each project. The second objective function assesses the risks inherent in the projects portfolio. For this, the portfolio benefit variance is estimated. The model takes into account restrictions on the number of available specific resources. The task appertains to multicriteria problems of Boolean programming. This model does not take into account the time factor and the distribution of projects in time.

The work [8] analyses the dynamic task of resource allocation between activities. There are three target functions proposed: the sum of incomes by business directions, the resulting environmental pollution, and entropy risk assessment. The problem is solved for a given planned period. Income from the business directions is ascertained using the production function, parameters of which are estimated in the prehistory. Resources between activities are allocated as a share of existent resources. The problem is solved by the method of concession.

The [9] proposes a mathematical model of the problem of optimizing the project portfolio, taking into account the start time of projects. The objective function is NPV for the entire portfolio. Restrictions take into account the availability of different resources types and the sequence of operations in projects fulfillment.

In the sources [10, 11] authors propose the mathematical model for projects port-folio optimization for the planned period. Each project is evaluated by the number of criteria, then the value of the generalized criterion is ascertained for it. The objective function of the optimization problem equals the sum of the generalized criteria values for all projects included in the portfolio. Each project can start in any year of the planned period. The model takes into account the restrictions on income from a portfolio, on the benefit earned, the investment resources, and the level of resource utilization in the projects portfolio management. The authors proposed the versions of the model in the deterministic and fuzzy formulation. The drawback of this model is that the task of maximizing the profits from the project portfolio for the planned period is not being solved explicitly.

The analysis of the literature on the optimization of projects portfolios showed that existing papers did not overlook mathematical models, which would count the dynamics of income and expenditure of funds in projects for the planned period, simultaneously would regard the social effect of the portfolio, the inherent risks and would impose restrictions on the sequence of projects fulfillment.

The purpose of the article is formulated as follows: to create a mathematical model of the projects portfolio optimization for the planned period, which would reckon the income and expenditure of funds over time, the social effect of the portfolio, the risks associated with its implementation, the necessary sequence of projects, and the mandatory inclusion of some project in the portfolio.

2   Results

It is considered the task of the portfolio optimization which consists of the projects able to start at the time segment [1, T]. As the time unit, we will understand here the segment of time in accordance with which the projects in the company and the flow of funds for them are planned. With regard to IT projects, it is convenient to choose one week as a unit of time, the duration of the sprint (2–4 weeks) or 1 month.

There are J projects under consideration, that can potentially be included in the portfolio or are being already fulfilled by the company. Project j can be started on periods $t = \overline{1, T}$, the payment for work from the clients can come on periods $t = \overline{1, T + g - 1}$, where $g = \max_{j=\overline{1,J}} l^{(j)}$, $l^{(j)}$ is the number of time units, during which there are being done the works on the j-th project and their financing. For the j-th project, the client will pay c_{jr} funds in the r-th period from its start, $r = \overline{1, l^{(j)}}$.

The j-th project expenses equal w_{jr} in the r-th period from the beginning of its realization, $r = \overline{1, l^{(j)}}$, where $l^{(j)}$ stands for the overall project implementation time (quantity of periods), during which the funds can be spent on it.

It is necessary to optimize the projects portfolio so that in every period $t = \overline{1, T + g - 1}$ there were enough funds for its realization, the sequence of interrelated projects was kept, every project would be realized no more than once. Herewith it is necessary to maximize the company's profit and the social effect from the portfolio's projects commitment and minimize the inherent risks.

$$\sum_{t=1}^{T}\sum_{j=1}^{J}\sum_{r=1}^{l^{(j)}} \left(c_{jr} - w_{jr}\right)x_{jt} \rightarrow \max, \tag{1}$$

$$\sum_{t=1}^{T}\sum_{j=1}^{J} R_j x_{jt} \rightarrow \min, \tag{2}$$

$$\sum_{t=1}^{T}\sum_{j=1}^{J} S_j x_{jt} \rightarrow \max, \tag{3}$$

$$\sum_{r=1}^{t} C_r^0 + \sum_{j=1}^{J}\sum_{p=1}^{t}\sum_{r=1}^{t-p+1} c_{jr}x_{jp} \geq \sum_{j=1}^{J}\sum_{p=1}^{t}\sum_{r=1}^{t-p+1} w_{jr}x_{jp},$$

for $t = \overline{1, T}$,

$$\sum_{r=1}^{t} C_r^0 + \sum_{j=1}^{J}\sum_{p=1}^{T}\sum_{r=1}^{t-p+1} c_{jr}x_{jp} \geq \sum_{j=1}^{J}\sum_{p=1}^{T}\sum_{r=1}^{t-p+1} w_{jr}x_{jp}, \tag{4}$$

for $t = \overline{T+1, T+g-1}$, if $T+g-1 \geq T+1$,

$$\sum_{t=1}^{T} x_{jt} \leq 1, j = \overline{1, J}, \tag{5}$$

$$x_{jt} \cdot \text{card}\, P_j - \sum_{p \in P_j}\sum_{m=1}^{t-l^{(p)}} x_{pm} \leq 0, \quad t = \overline{1, T}, \tag{6}$$

$$\sum_{t=t_{s1}}^{t_{s2}} x_{st} = 1, \tag{7}$$

$$x_{jt} \in \{0, 1\}, \quad j = \overline{1, J}, t = \overline{1, T}, \tag{8}$$

where R_j represents risks, related to the j-th project realization,

S_j – the social effect from the j-th project,

C_r^0, $r = \overline{1, T+g-1}$, the funds, that company can spend on the projects portfolio realization in the r period. These funds can have a negative value if the company needs to gain funds by realizing the project in r period,

x_{jt} is a boolean variable equal to 1 if the j-th project is started in the t-th year, and equal to 0 otherwise.

For the j project there can be specify a set P_j of numbers of projects, that must be committed before the project j start.

The objective function (1) of the task presents the residual between the all-projects revenues and expenses, starting from the first period to the T-th. As the j-th project is being fulfilled within the $l^{(j)}$ periods, $j = \overline{1,J}$, the objective function (1) counts revenues and expenses for projects starting from the first period to the T-th either after the T period, during $g - 1$ periods, $g = \max\limits_{j=\overline{1,J}} l^{(j)}$.

The objective function (2) reflects the risks associated with the realization of the j-th project. For each project, a list of the most significant risks is compiled. Risks are estimated by multiplying the probability of a risk event occurring on the significance of its consequences. The consequences of a risk event are scored from 0 to 10 in accordance with Table 1.

Table 1. Estimates of the risk events consequences

Negative consequences	Points
Impacts resulting in the termination or complete failure of the project	10
Impacts that lead to the very significant project delays, budget overruns, deterioration of project product quality	8–9
Impacts that lead to significant project delays, budget overruns, deterioration in project product quality	6–7
Impacts that lead to not very significant project delays, budget overruns, deterioration in project product quality	4–5
Impacts that lead to insignificant project delays, budget overruns, deterioration in project product quality	2–3
Negative effects are almost not noticeable	1
No negative effects	0

The j-th project risks assessment equals:

$$R_j = \sum_{k=1}^{K^{(j)}} \alpha_{kj} e_{kj},$$

where α_{kj} represents the probability of the k-th risk event for the j-th project,

e_{kj}– the negative consequences (measured in points) of the k-th risk event for the j-th project,

$K^{(j)}$– the quantity of the risk events for the j-th project.

The social effect of the j-th project fulfillment may consist of personnel's qualification increase because of this project, of raising staff salaries, of solving social problems of the team or community that this project affects. Table 2 shows the social effect expressed in points.

The constraint (4) requires that in the period t, $t = \overline{1, T+g-1}$, the funds earned in projects starting before and within the t period were more or equal to the expenses in these periods.

The constraint (4) when $t = \overline{1,T}$ applies to the periods when the fulfillment of projects included in the portfolio can be started.

Table 2. Estimates of the project social effect

Social effect	Points
Very significant social effect	9–10
Significant social effect	7–8
Middle-sized social effect	5–6
Insignificant social effect	3–4
Social effect is almost not noticeable	1–2
No social effect	0

The constraint (4) when $t = \overline{T+1, T+g-1}$, applies to the periods after T, within which there continues the fulfillment of projects started before and during T, inclusively.

The first summand in the left part of the constraint (4), either $t = \overline{1, T}$, or $t = \overline{T+1, T+g-1}$, i.e.,

$$F_1 = \sum_{r=1}^{t} C_r^0$$

are the accumulated to and within the t period funds, which the company can spend on the projects portfolio fulfillment.

The second summand in the left part of the constraint (4) for $t = \overline{1, T}$, i.e.

$$F_2 = \sum_{j=1}^{J} \sum_{p=1}^{t} \sum_{r=1}^{t-p+1} c_{jr} x_{jp}$$

represents the funds accumulated to and within the t period from projects, which were started during the periods from the 1-st to the t-th, inclusively.

The right part of the constraint (4) for $t = \overline{1, T}$

$$F_3 = \sum_{j=1}^{J} \sum_{p=1}^{t} \sum_{r=1}^{t-p+1} w_{jr} x_{jp}$$

are the expenses accumulated to and within the t period from projects which were started during the periods from the 1-st to the t-th, inclusively.

The second summand in the left part of the constraint (4) for $t = \overline{T+1, T+g-1}$, i.e.

$$F_4 = \sum_{j=1}^{J} \sum_{p=1}^{T} \sum_{r=1}^{t-p+1} c_{jr} x_{jp}$$

represents the funds accumulated to and within the t period from projects, which were started from the 1-st to the T-th periods, inclusively.

The right part of the constraint (4) for $t = \overline{T+1, T+g-1}$, i.e.

$$F_5 = \sum_{j=1}^{J} \sum_{p=1}^{T} \sum_{r=1}^{t-p+1} w_{jr} x_{jp}$$

are the expenses accumulated to and within the t period from projects which were started during the periods from the 1-st to the T-th, inclusively.

The constraint (5) requires that any j-th project, when $j = \overline{1, J}$, was fulfilled no more than once.

The constraint (6) suggests that, before the j-th project start, the projects from the P_j set must be fulfilled.

The constraint (7) allows specifying the requirement of the mandatory inclusion in the portfolio of the s-th project on the time segment $[t_{s1}, t_{s2}]$.

The problem (1)–(8) pertains to the multicriteria dynamic Boolean programming problems. We can also characterize the (1)–(8) problem as the multicriteria non-Markov's dynamic problem of discrete optimization [12]. The term "non-Markov's" refers to the optimization problems, where the object state at the t-th stage equals to the function of state at the previous $t-1$ stage and controls at the stages t, $t-1$, $t-2$,..., $t-p + 1$. I.e. the aftereffect of the previously applied controls is taken into account. Therefore, for the projects portfolio optimization problem, the decision about the start of certain j-th project $l^{(j)}$ periods in duration, made within the t period, will affect the portfolio state during the $t+1, t+2, \ldots, t+l^{(j)} - 1$ periods.

3 Conclusion

The works in the field of projects portfolio optimization have been analyzed. The aim of the paper has been to create the mathematical model, which would develop the known approaches facilities. The present paper suggests the dynamic model of the projects portfolio optimization problem. The model reckons the incomes and expenses during the projects fulfillment, the existence of funds for the portfolio realization in every given period, the social effect of portfolio implementation, its inherent risks, constraints for the projects realization sequence, and the mandatory inclusion of some project in the portfolio on the time segment. The regarded problem appears to be the non-Markov's, or the problem with the aftereffect of the decisions made at previous stages. In prospect we presume the creation of the method of this problem solving, that will take into consideration its specific features.

References

1. Liagkouras, K.: Novel multiobjective evolutionary algorithm approaches with application in the constrained portfolio optimization. A dissertation for the degree of Doctor of Philosophy, University of Piraeus (2015)
2. Elbok, G., Berrado, A.: Towards an effective project portfolio selection process. In: Proceedings of the International Conference on Industrial Engineering and Operations Management, Rabat, Morocco, pp. 2158–2169, 11–13 April 2017

3. Ghasemzadeh, F., Archer, N.P.: Project portfolio selection through decision support. Decis. Support Syst. **29**(1), 73–88 (2000). https://doi.org/10.1016/S0167-9236(00)00065-8
4. Stummer, C., Heidenberger, K.: Interactive R&D portfolio analysis with project interdependencies and time profiles of multiple objectives. IEEE Trans. Eng. Manage. **50**(2), 175–183 (2003). https://doi.org/10.1109/TEM.2003.810819
5. Kremmel, T., Kubalik, J., Biffl, S.: Software project portfolio optimization with advanced multiobjective evolutionary algorithms. Appl. Soft Comput. **11**(1), 1416–1426 (2011). https://doi.org/10.1016/j.asoc.2010.04.013
6. Morris, P., Pinto, J.K.: The Wiley Guide to Project, Program, and Portfolio Management. Wiley, Hoboken (2010)
7. Rădulescu, M., Rădulescu, C.Z.: Project portfolio selection models and decision support. Stud. Inform. Control **10**(4), 275–286 (2001)
8. Barkalov, S.A., Bakunets, O.N., Gureyeva, I.V., Kolpachev, V.N., Russman, I.B.: Optimizatsionnyye modeli raspredeleniya investitsiy na predpriyatii po vidam deyatelnosti. IPU RAN, Moscow (2002)
9. Shou, Y., Huang, Y.: Combinatorial auction algorithm for project portfolio selection and scheduling to maximize the net present value. J. Zhejiang Univ. Sci. C Comput. Electron. **11**(7), 562–574 (2010). https://doi.org/10.1631/jzus.c0910479
10. Kononenko, I.V., Bukrieieva, K.S.: Model i metod optimizatsii portfeley proyektov predpriyatiya dlya planovogo perioda. Vostochno-Evropeyskiy zhurnal peredovykh tekhnologiy. **43**(1/2), 9–11 (2010)
11. Kononenko, I.V., Bukrieieva, K.S.: Metod formirovaniya portfelya proyektov predpriyatiya dlya planovogo perioda pri nechetkikh iskhodnykh dannykh. Upravlinnia rozvytkom skladnykh system. Zbirnyk naukovykh prats. **7**, 39–43 (2011)
12. Richter, K.: Dinamicheskiye zadachi diskretnoy optimizatsii. Radio i svyaz, Moscow (1985)

Agrarian Enterprises Mathematical Ontological Model Development

Iana Savitskaya⬥, Victor Smolii(✉)⬥, and Vitalii Shelestovskii

National University of Life and Environmental Science of Ukraine,
Kiev 03041, Ukraine
yasawitskaya@gmail.com, dr.v.smoliy@gmail.com,
Shelestovskiyvit@gmail.com

Abstract. Food sustainability is a one of the most-important problems of humans for today and tomorrow. Information and Communication Technologies integrating in Agriculture can solve some problems in this direction. This way based on such newest technologies as IoT, Cloud Computing, AI and others. But this technologies not guarantying any positive result without according business logic models and Agro knowledge gathering and sharing.

An aim of this paper to justify a choice of tools and methods for design a data and knowledge models for agricultural enterprise control and management systems implementation. This solution must to solve main targets formulated by the Common Agricultural Policy of EU.

An attention focused on ontology technology of knowledge presenting and collecting. In the work is proposed a simple example of farm enterprise ontology, which be used like a base for developing of corresponding control system. Also proposed a way for advance abilities and usability of this ontology model.

Keywords: Ontology · OWL · RDF · Graph · Management system · CAP

1 Introduction

According to the CAP – Common Agricultural Policy [1], the agriculture digitalization is defined by Euro-Commission as a priority. The analysis of literary sources points not only to the big prospects, but also to a number of difficulties. Thus, the growth of the computerized agrarian sector in Germany amounted to about 12% [2] in 2017. Big companies principal direction of equipment development that is intelligent and easy to integrate into enterprise management system [2]: "The priority no longer lies in optimizing machines and increasing their drive power. ... more interested in looking for possibilities to make processes connected and intelligent".

Problems of CAP implementing [2] are the lack of appropriate infrastructure and management. This leads to the fact that only 12% of farms in Germany in 2016 used the farm- management systems, with the fact that more than half use digital technology.

© Springer Nature Switzerland AG 2020
A. Palagin et al. (Eds.): MODS 2019, AISC 1019, pp. 243–248, 2020.
https://doi.org/10.1007/978-3-030-25741-5_24

2   Structure and Functions of a Farm Management System

2.1   Main Goals of Farm Management System Development

According to AGENSO (AGricultural and ENvironmental SOlutions) [3], the main areas of agrarian sector digitalization with the European Commission support are:

- Farming solutions – Support decision for a more robust farm management and crop growing control;
- Environmental solutions – for protecting biodiversity and ecosystems taking care of the environment;
- Water management;
- ICT by IoT for agricultural and advanced spatial mapping techniques for data visualization;
- Services like a Precision Agriculture or Environmental and Research services.

The development of these areas should lead to the new SMART Farming technology [4], which combines such advanced ICT (Information and Communication Technologies) solutions as precision (in agriculture context) devices, IoT, geo-positioning systems, touch-sensitive and executing devices, BigData technologies, mobile devices without humans (Unmanned Aerial Vehicles - UAVs, drones), robotic systems, and so on. But according to the same source, these systems represent such more classic technologies as:

- Management Information Systems;
- Precision Agriculture;
- Agricultural automation and robotics.

These systems and technologies should combine and complement each other at different functional levels of the agrarian enterprise (according to the preceding paragraphs):

- General and operational management, control and planning;
- Collection, systematization and processing on the basis of intelligent systems (including Decision Support Systems) in order to generate technological operations;
- Automated technological operations execution accordingly to the solutions found.

Taking into account that most of the digitalization issues in agriculture were formulated and partly solved under the Horizont 2020 program, the European Commission decide to continue working in this direction after 2025 too. The main working areas are [5]:

- Access to Smart Agriculture for all farms;
- Modern and simple support for farm investment in the future of the CAP;
- Setting the stage for advisory services of the future;
- Demonstration and sharing of topical knowledge;
- The Review and Update of Educational curricula;
- Ensuring rural broadband connectivity;
- How to Simplify, Innovate and Network the Funding Instruments.

It solves three main problems [1]:

- Production of a huge amount of high-quality safe products is ensured;
- Raising standards and quality of life in rural areas;
- Sustainable and weather-smart production is ensured.

2.2 An Existing Systems Analyze

Examples of systems developed with the European Commission support and covering some of the challenges are - AgriOpenData [6], SITI4Farmer, SITI4Data, SITI4Knowlidge [7], AgroAPI [8] and some another systems with a similar functional set or narrowly specialized.

AgriOpenData tracks the conditions for plants development and their production and allows according to the Blockchain technology quality confirmation and some organizational processes automation.

ABACO Groupe products [7], integrated to one networked information resource, allow to solve task complexes associated with a variety of processes and grouped into three frameworks by appointment. In order to solve problems related to agriculture, 18 tools are offered, and in the SITI4Farmer package only 5 of them are presented: the Decision Support System (DSS), the Agro-Weather Data and Indices module, the module Management Support (SITI4Farmer), Mobile Support Module (SITI4Fields), and Custom Apps.

AgroAPI solves a narrow range of tasks – provides information from satellite systems in the form of images for further territory and plants characteristics analysis.

The FASAL microcontroller system [9] is not a European product – it is developed by Indian engineers and scientists and helps small farms to control the state of plants and regulate the conditions of their cultivation.

All European systems use both IoT and Cloud Computing technologies as tools for data collecting, processing, services access organization and mobile devices for organizing user interfaces.

But these are closed systems that do not solve, for example, such tasks as the accumulation knowledge and sharing of them, the construction of a basis for counseling and education.

2.3 Tools and Background for Models Developing

Obviously, the difficulties of these tasks realizing are related to the agrarian objects appropriate models development, methods and tools for intelligent processing of received data, the allocation of knowledge, their accumulation and distribution.

It should be noted, that technologies to solve some of these problems for today there are.

First and foremost, these are Knowledge Bases and their corresponding tools. Taking into account that the agrarian management system combines many diverse and heterogeneous information, it is logical to conclude that it is a heterogeneous system with knowledge that is difficult to describe by a framing or production model and, consequently, rational use of network (semantic) models.

According to this fact, it is rational to choose a toolkit that allows to describe such graph systems. For today, these technologies are ontology [10] and semantic web [11], and tools such as OWL [12] and RDF [13].

Let us consider the possibilities and features of modeling based on simplified information structures of agricultural enterprise management system, since a detailed review is difficult within the short work. Ontology is a hierarchical graph structure, in the agrarian enterprise model can contain concepts (classes), shown on the Fig. 1.

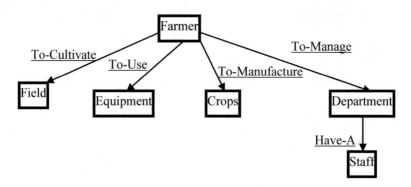

Fig. 1. A simple ontological model of an agricultural enterprise

Systems used by farmers can vary in scale from FASAL-like ones to AgriOpenData-like. But ideally, they need to be integrated into a general knowledge-gathering system. That's why a model that represents knowledge to classes, slots, and instances of the subject area should be built.

To the main information classes of the easiest representation level given by a farmer enterprise are included: crop, field, soil, microclimate, growth stage, agro technological operations and devices and so on with the corresponding connections (see Fig. 2).

The main areas of information at this level of enterprise representation can be considered the concepts "Field" and "Crop".

The presented ontology does not contain a description of slots and instances of classes for the perception simplifying purpose. For example, the soil parameters that need to be monitored and taken into account are divided into 2 groups – basic and additional.

The main ones include those that can be obtained using modern precision sensors [14] - humidity, salinity, temperature, pH, nitrogen, potassium, phosphorus, CO_2, O_2. Additional parameters are soil parameters that are also important, but can not be determined quickly: soil type, soil structure, biological content and background, underground water level, and so on.

Taking into account these characteristics leads to the choice of plants that can grow on a certain type of soil, and thus will determine the instance of the class "optimal soil" for each instance of the "Crop" class.

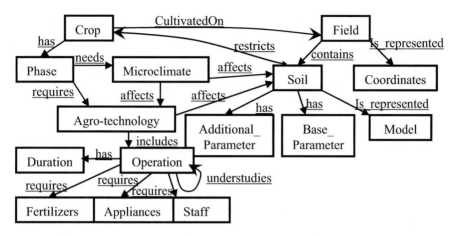

Fig. 2. A simple ontology of a crop growth

2.4 A Way to Advance of a Classical Ontology Model

From mathematical point of view, ontology represents, as indicated earlier, the graph. It means that the apparatus of graphs theory can be used for analysis and search of knowledge. In the basic version, the ontology corresponds to an oriented graph that defines hierarchy of classes.

As can be, the ontology can contain cycles (see Fig. 2) and bilateral relations (not shown) – for example, the soils influence to the microclimate in the spatial point, where they are located. On the one hand, this complicates the model presentation, but from another hand it gives additional opportunities. The semantic network obtained during the ontology construction can be considered, for example, as a technological map of agrarian enterprise management process.

By adding weight ratios to the edges in the graph, it is possible to advance system abilities. In this case, some manufacturing problems can be solved by using well-known standard mathematical methods, for example, management processes problem optimizing, composition and timing of agricultural operations determination for transition from one crop cultivation to another, operations set definition for implementation of agro-technical action in conditions of limited resources.

The graph theory assumes each vertex $v_i \in |V_N|$, $i = 1..N$ of the labeled graph $G(N)$ to have a weighting factor v_i that corresponds to "importance", and $d_{i,j}$ - the "weight" of the arc $e = \{V_i, V_j\}$ is to determine the distance between the vertices it connects. The weighted distance between two vertices is defined as $l_{i,j} = v_i{\cdot}d_{i,j}$.

Then, for example, the task of optimizing the enterprise management process will be to obtain a graph, with a minimum sum of weighted distances:

$$G(|V_N|, |E_N|) \rightarrow G(|V_M|, |E_M|) \Big|_{\sum v_k \cdot d_{k,s} \rightarrow \min}, \text{ where} \tag{1}$$

$$|V_M| \in |V_N|, |E_M| \in |E_N|$$

However, in the ontological graph, such a classic approach is difficult to apply, since vertices can have a different nature. For example, Field is combined with Soil and Coordinates, between which comparative weights cannot be entered, that is, the classical markup approach can be used for vertices of the only one class objects. Accordingly, there also arises the problem of developing methods for calculating heterogeneous ontological graphs.

References

1. The common agricultural policy at a glance. https://ec.europa.eu/info/food-farming-fisheries/key-policies/common-agricultural-policy/cap-glance_en. Accessed 24 April 2019
2. Digitalising agriculture: Opportunities and market control. https://www.euractiv.com/section/agriculture-food/news/digitalising-agriculture-opportunities-and-market-control/. Accessed 24 April 2019
3. AGENSO. http://www.agenso.gr. Accessed 24 April 2019
4. What is Smart Farming? https://www.smart-akis.com/index.php/network/what-is-smart-farming/. Accessed 24 April 2019
5. Smart-AKIS Policy Report and Briefs for mainstreaming Smart Farming in the new CAP. https://www.smart-akis.com/index.php/2018/08/29/smart-akis-policy-report-policy-briefs-smart-farming-europe/. Accessed 24 April 2019
6. AgriOpenData. https://www.agriopendata.it/. Accessed 24 April 2019
7. ABACO Group. https://www.abacogroup.eu/. Accessed 24 April 2019
8. AgroAPI. https://agromonitoring.com/. Accessed 24 April 2019
9. FASAL. https://fasal.co/. Accessed 24 April 2019
10. Gruber, T.: Toward principles for the design of ontologies used for knowledge sharing. Int. J. Hum.-Comput. Stud. **43**(5–6), 907–928 (1995). https://doi.org/10.1006/ijhc.1995.1081
11. Dustdar, S., Falchuk, B.: Semantic web. In: Furht, B. (ed.) Encyclopedia of Multimedia. Springer, Boston (2006). https://doi.org/10.1007/0-387-30038-4
12. The W3C Web Ontology Language (OWL) is a Semantic Web language designed to represent rich and complex knowledge about things, groups of things, and relations between things. https://www.w3.org/2001/sw/wiki/OWL. Accessed 24 April 2019
13. Resource Description Framework (RDF). https://www.w3.org/2001/sw/wiki/RDF. Accessed 24 April 2019
14. Tetralytic Soil Probe. https://order.teralytic.com/products/soil-probe. Accessed 24 April 2019

The Model of Information Security Culture Level Estimation of Organization

Serhiy Shkarlet⬤, Vitalii Lytvynov⬤, Mariia Dorosh$^{(\boxtimes)}$⬤,
Elena Trunova⬤, and Mariia Voitsekhovska⬤

Chernihiv National University of Technology,
Shevchenko Street 95, Chernihiv 14035, Ukraine
rector@stu.cn.ua, v.v.lytvynov.dept@gmail.com,
mariyaya5536@gmail.com, e.trunova@gmail.com,
m.voitsekhovska@gmail.com

Abstract. Deep penetration of computer networks into all spheres of organization's life requires new approaches to ensuring security for survival in a context of continuous renewal of cyber threats. Since the object of information operation is a person, there is a need for the developing of information systems that are aimed not only at the technical protection of computer systems and technology, but also to prevent the impact of social engineering methods.

In order to reduce such influences, it is necessary to work continuously to increase the level of information security culture of the organization. The model of the system of complex assessment of the information security culture (ISC) level of personnel as an important component of the overall information security organization presented in the article contains six levels and is divided into two subsystems. The first subsystem includes ranking of factors, and the second - the definition of the integral indicator of the personnel ISC level of the organization. The peculiarity of the proposed model lies in the use of fuzzy decision-making methods that allow to simulate the change in the properties of the object, expressed in the form of qualitative relationships. Also, the model takes into account the assessment of the security specialists competence, depending on the role they play at different levels of its provision.

The model can be used in the development implementation and audit, of integrated information security systems.

Keywords: Information Security Culture · Personal culture ·
Organizational culture · Corporate culture · Competence · Role

1 Introduction

In today's global networked organizations risk management and information asset protection have become critical components of any successful IT strategy. Wherever the threat came from technology and politics BYOD (Bring Your Own Device) or so-called "Complex permanent threats", which implies long-term unauthorized use of third-party information resources for theft or espionage, – network security is becoming a critical factor that enhances business efficiency and the social value of networked information systems [1]. On the other hand, the problem of Information Operations by

© Springer Nature Switzerland AG 2020
A. Palagin et al. (Eds.): MODS 2019, AISC 1019, pp. 249–258, 2020.
https://doi.org/10.1007/978-3-030-25741-5_25

means of not only by fake news is not a new problem and its spread in networks is well studied. In addition, worthy of attention fake mail, fake links, fake advises or help (technical, medicine, law) which people are used to research, study, business or work. Today, the danger comes from not only news or advertising but also covers various aspects of human life.

Since the person became the object of Information Operations, to reduce the impact a systematic approach is required. It must contain elements of various technical, economic, philosophical, and sociological, modern developments. It is important to understand not only the technical aspects of computer networks building and defensing but also the ability to model the response and decision-making by a person as a result of Information Operations in order to create effective information systems of protection and prevention of attacks. Today, for solving these issues, an expert systems and systems based on neural networks using different algorithms for building knowledge bases in technical information security systems are actively used.

Also, all over the world, to search for and collect information in computer networks, special monitoring data collection systems are used. Such systems automatically intercept any information being monitored as soon as it appears on visible web.

At the same time, the role of user's ISC is growing as a key factor in reducing the impact of the human factor on general security is gaining ever-greater significance.

Therefore, it is clear that the information security culture is a complex feature of the organization's information security that reflects the system of indicators for ensuring technological processes and the status of training personnel that correspond to the permissible risks.

In our study we definite the personal Information Security Culture as a set of characteristics, that contains personal competencies, attitudes, assumptions, perceptions, and values of employees in the InfoSec domain during interaction in organization's information environment, including its development and improving.

2 The Aim of Research

This research is devoted to the decision of the question of the automated system modeling for the level of employees' ISC assessment as an important component of the complex organization security.

Also, the important issue of this study is to justify the need to create information systems for managing the ISC and integrate them into existing technical systems to provide integrated information security in the organization.

3 Related Works

In the current situation, one of the driving forces of a successful business activity is technology, mobility, and accessibility of services. Investments in information assets are valued on a par with tangible capital, and protection of these information assets deservedly require increased attention. Despite the intensive development of technical information defense tools, which the owners and persons responsible for information assets rely in most cases, the employees are not the last factor affecting the security and integrity of the organization's information assets. The security of resources, computing power, software and end users still depends on the ability of staff to withstand external and internal information threats, the ability to recognize and prevent the occurrence of a critical situation in a timely manner.

In this connection, scientists and IS-specialists regularly point out the need to influence the information security of organizations through the formation of an adequate level of information security culture. The culture phenomenon is using long and effectively to form the required secure behavior and interactions, the level of awareness in critical cases and the safe labor activity provision.

The problems of forming the required level of ISC for employees of the organization over the past decade are rightly present among the works devoted to the protection of information assets and the computer networks security.

Van Niekerk and Von Solms [2] emphasize the importance of employees ISC introducing as a subculture within organization in order to manage the influence of human factor on the organization's information security. Earlier, Ngo et al. indicated in paper [3] the inseparability of information security culture from corporate culture.

Need to define the level of security culture of organization to take steps for improving the current secure situation highlighted by Furnell and Thomson [4]. The mentioned research allows us to apply the concept of natural employee behavior to form the information security culture.

Intensive research activity of Da Veiga et al. resulted in a number of papers, devoted to the influence of human factor on information security management system (ISMS). In particular, in the paper [5] notes the impact of the employees ISC on mitigating the information security risks associated with the employees and assets interaction.

Increasingly, researches on the ISC take into account such factors as national mentality and culture. An example is the study of Flores, Antonsen and Ekstedt [6] that indicates the impact of the above factors on the organizations' activities in a global environment.

This section will not be complete without mentioning the names of Schlienger and Teufel, whose work [7] was the starting point for many researchers and the guidance to action for company executives interested in further enhancing the protection of information assets by employees.

Taking into account obstacles while instilling and developing employees ISC leaders can face with (as fear, resistance and confusion), Alhogail and Mirza proposed an effective change management models and multistep framework described in [8].

In addition, a large and laborious work on the analysis and systematization of research in the field of information security culture of organizations was done by Sherif, Furnell and Clarke, and presented in the paper [9].

The above works are only a small part of modern research based on a deep understanding of human psychology, social interactions, personnel management, information and network security, and corporate culture. New information technologies appear and develop every day, and traditionally information security will challenge.

However, the issue of automation of ISC management is still insufficiently studied. There are almost no comprehensive software tools that allow to evaluate, plan, and control changes in the organization's information security culture. Such tools should take into account the technical and personal components of the information security culture, as well as the degree of risk at different levels of computer networks organization.

4   Main Methods and Research

Modeling a system for assessing the ISC of the organization occurs with the involvement of a set of methods of system analysis such as impregnation and clustering, and includes elements of fuzzy logic. The combination of these approaches in a single information system provides the opportunity to integrate different assessment systems, and reduce the risks of subjectivity of the resulting evaluation.

The effectiveness of the implemented InfoSec system depends on the personnel, especially in such points as the need to comply with the IS requirements, the responsibility, the understanding of the potential consequences of incidents and the perception of the IS monitoring as shown on Fig. 1.

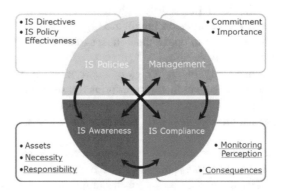

Fig. 1. The model of organization's ISC lever evaluation

So in spite of the rapid development of technical systems for IS it is still almost impossible to exclude a person from the process of creating, processing, transmitting information, which maintains a high degree of vulnerability to attacks based on social engineering.

As part of the organization's activities, each of the employees performs certain job functions, including in the information security domain. Depending on this, certain competency requirements are formed, ensuring certain InfoSec level in the process of functional duties performing.

At the beginning of the organization's personnel ISC level estimation, we can use the questionnaire. The questions concern personal information security aspects and the knowledge and experience in the field of information security. The source for the formation of questions can be the international standard ISO/IEC 27001. Taking into account the activities of the organization and its employees, the ISO/IEC 27032 standard may also be involved.

Set of questions in the subject area forms the database that to one degree or another relate to information security.

Hierarchical six-level fuzzy model of organization's personnel ISC level estimation is divided in two subsystems, each of them has its own separate functions, namely: subsystem 1 – factors ranking (1st and 2nd levels), and subsystem 2 – determination of the integral indicator of the organization's ISC level (3rd, 4–6th levels), as it shown on Fig. 2.

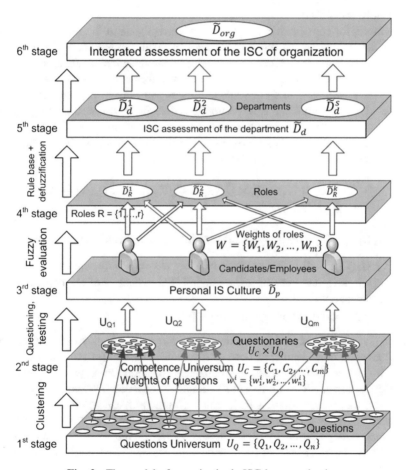

Fig. 2. The model of organization's ISC lever evaluation

It should be noted that the system is cyclical, since the process of raising the ISC level is continuous, and the assessment of the ISC level must take place within the audit of organization's information security. The need for a regular InfoSec audit is continuous assess the real status of the security of IS and/or ITS resources and their ability to withstand external and internal threats to information security that are constantly changing and adapting.

At the same time, the InfoSec audit is a systematic process for obtaining objective quantitative and qualitative assessments of the current state of security of the information system. It is carried out taking into account three main factors: personnel, processes and technologies.

The assessment of the organization's information security processes and technologies is more or less defined and provided by various methods, and the question of a comprehensive assessment of the level of culture of information security of the personnel, as an important component of the overall security of the organization, remains insufficiently researched.

Therefore, for comparative analysis of the current state of the information system, which is determined by the results of the survey, with the test model of the requirements of the standard ISO 27001, and/or with its previous indicators, an effective tool is required that will be supported by the relevant information technology.

To do this, you need to form a single integral indicator of the organization's ISC D_{org}. To solve the problem, it is proposed to use fuzzy decision-making methods, which allow simulating a smooth change in the properties of an object, as well as unknown functional dependencies expressed in the form of qualitative connections.

So, consider the mathematical display:

$$f : C_1 \times C_2 \times \ldots \times C_m \times \tilde{Q}_1(T) \times \tilde{Q}_2(T) \times \ldots \times \tilde{Q}_n(T) \to \tilde{D}_{org}$$
$$= \{\tilde{D}_d^1, \tilde{D}_d^2, \ldots, \tilde{D}_d^s\}, \tag{1}$$

where $C_1, C_2, \ldots, C_m$ – an array of information security personnel competencies;

$\tilde{Q}_1(T), \tilde{Q}_2(T), \ldots, \tilde{Q}_n(T)$ – an array of fuzzy answers to questions;

$\tilde{D}_d^1, \tilde{D}_d^2, \ldots, \tilde{D}_d^s$ – an array of fuzzy integral ISC indicators of the departments and the organization;

$\tilde{D}_{org}$ – fuzzy indicator of organization's ISC.

Note that when constructing a model for the formation of a linguistic assessment of the ISC level as input variables, both as quantitative factors (m – number of competencies; n – number of questions; r – number of positions/employees; s – number of departments), and qualitative ones $Q_i(T)$ – linguistic questions evaluation ($j = 1, \ldots, n$) with a term set $\{poor, normal, good\}$, and/or $\{low, average, high\}$ are used.

Systems of fuzzy products are carried out in accordance with the method of fuzzy logic conclusion proposed by Bellman and Zadeh in the work [10]. In our view, evaluating of the ISC level can be provide with use MATLAB package Fuzzy Logic Designer.

The First Stage in presented model is forming a number of questions.

An array of factors (competences) $U_C = \{C_1, C_2, \ldots, C_m\}$ is forming, which contain security competencies and an array of indicators (a set of questions) $U_Q = \{Q_1, Q_2, \ldots, Q_n\}$, that characterize the object of research (ISC). Note that competence includes knowledge and understanding of how to act, and of how to be. The subject area in which the individual is well-informed, and in which he is ready to perform activities in safety.

In turn, the assessment of competencies can be carried out through a questionnaire (testing), reflecting the employee's security activities. These competencies can be defined in the requirements for the qualification of specialists according to their specialty, and may also be formed in separate workplaces depending on the roles performed within the framework of the general security system.

Formation of the array of competencies and questions can be conducted by two (different) groups of users – engineers of the organization's information security, heads of security departments or leaders of the organization.

The Second Stage. Criterion sets (clusters) $C^i : Q^i \subset U_Q$ that characterize separate competencies for the corresponding sets of questions are determined. In this case, within the limits of each separate competence, the set of questions has qualitatively homogeneous indicators, which are formed on the basis of the kind of professional activity of a specialist.

Given the conditions for the formation of clusters, Fuzzy Clustering seems to be the most appropriate clustering method. This method allows including the same object (a question that is a variable) to similar competencies. According to the results of clustering, questionnaires or tests are prepared for specialists' interviewing.

Within each competence set $C^i : Q^i \subset U_Q$, the task is to determine the relative values of the issues that form it. The purpose of this task is to determine the weight of their influence on the elements of the following levels of the hierarchy.

It is expedient to implement the implementation of the Saati method [11] by pairwise comparing questions on the appropriate scale. Thus, for each competence C^i from the set $= \{C_1, C_2, \ldots, C_m\}$, a matrix of pairwise comparisons of a plurality of questions $Q^i = \left(q_{ij}^i \right)$ will be formed, on the basis of which their relative weight vector $\{w_1^i, w_2^i, \ldots, w_n^i\}$ is determined. And the matrix of pairwise comparisons $C = c_{ij}$ of the set of competences, on the basis of which determines the weight of each competence $W = \{W_1, W_2, \ldots, W_m\}$.

The weight of each role P (the hierarchy of roles is known in advance) is determined by the use of the Fishburn's principle [12] according to the formula:

$$P_k = \frac{2 \cdot (r - k + 1)}{(r + 1) \cdot r}, \tag{2}$$

where P_k – weight coefficient of i-th role (post);

r – number of posts (roles);
k – serial number (rank) of the post.

To implement the second level, the neural network is also can be used.

The Third Stage. Based on the results of the 1st and 2nd levels, the personal integral level of the specialists' ISC is assessed.

Note that when determining the personal level of ISC, the weight of W_i of each i-th competence is considered equal, and equals $1/m$, in other words, the overall fuzzy assessment of the personal ISC level (degree) of specialist is determined by the formula:

$$\tilde{D}_p = \sum_{i=1}^{m} \sum_{j=1}^{n_j} \frac{1}{m} \cdot w_{ij} \cdot \tilde{Q}(T)_{ij}, \tag{3}$$

where m is a number of competencies;

n_i – number of questions in the i-th competence;
W_{ij} – the weight of j-th question belonging to the i-th competence;
$\tilde{Q}(T)_{ij}$ – linguistic evaluation of j-th question of i-th competence.

The Fourth Stage. Since different roles are being implemented at different levels of information security of the organization such as users, maintenance of security systems, comprehensive security assurance, security management, etc., requirements for their security competencies will be different.

Consequently, within the framework of the activities of the organization (department), each employee performs certain functions (roles) in the information security system, which are determined by the appropriate set of competencies $C = \{c^1, c^2, \ldots, c^m\}$. The training, experience and qualification of all individuals involved in any activities related to the full life cycle of InfoSec should be evaluated for a specific application. To assess the competence of individuals in the performance of their duties (roles), it is necessary for each post to form their own set of competencies $R^k : C^j \subset U_C$.

The overall fuzzy assessment of the ISC for a role $\tilde{D}_R$ (degree) is calculated by the formula:

$$\tilde{D}_R = \sum_{i=1}^{m} \sum_{j=1}^{n_i} W_i \cdot w_{ij} \cdot \tilde{Q}(T)_{ijk}, \tag{4}$$

where m is a number of competencies;

n_i – number of questions in the i-th competence;
W_i – weight of i-th competence;
w_{ij} – weight of j-th question belonging to the i-th competence;
$\tilde{Q}(T)_{ijk}$ – linguistic evaluation of j-th question of i-th competence for k-th role.

The Fifth Stage. The overall fuzzy assessment of the ISC for the department $\tilde{D}_d$ (degree) is calculated by the formula:

$$\tilde{D}_d = \sum_{i=1}^{r} P_k \cdot \tilde{D}_R^k, \tag{5}$$

where r – a number of posts (roles);

P_k – weight of i-th role (post);
$\tilde{D}_R^k$ – estimation (quantitative, qualitative) for k-th role.

The Sixth Stage. The integral indicator of the organization's ISC level $\tilde{D}_{org} = \{\tilde{D}_d^1, \tilde{D}_d^2, \ldots, \tilde{D}_d^s\}$ is defined as the geometric average of the integral indicators of the departments:

$$\tilde{D}_{org} = \sqrt[s]{\tilde{D}_d^1 \cdot \tilde{D}_d^2 \cdot \ldots \cdot \tilde{D}_d^s}. \tag{6}$$

Indicators $\tilde{D}, \tilde{D}_d, \tilde{D}_{org}$ can be considered as functions of time $\tilde{D}_R(t), \tilde{D}_d(t), \tilde{D}_{org}(t)$, namely, in dynamics. It provides an opportunity to analyze the time series of the corresponding functions obtained over a certain period of time. They characterize the change of ISC for a role, department, organization depending on actual learning outcomes and lessons learned, and their trends determine the direction of change. A trend analysis may prove to be very useful both for the organization as a whole and for employees in particular, and it allows us to conclude that the current and future status of the ISC and the effectiveness of the proposed training courses and other information security measures.

The general formula for determining the effectiveness of the proposed training courses and other measures to improve the InfoSec is:

$$E_{ff} = \frac{\tilde{D}(t_{curr})}{\tilde{D}(t_{prev})}. \tag{7}$$

The results of the calculations in the dynamics will reflect the effectiveness of the work of the security department with the personnel.

5   Conclusions and Perspectives

The developed system model of integrated assessment of the ISC level of employees of organization takes into account the different levels of assessment of the competence in the security domain, depending on the roles they play in the general organization's system of information security. The transition from qualitative to quantitative indicators is carried out through the use of fuzzy logic methods, and such transformations may take place at the last three levels of evaluation, if it is necessary to obtain indicators in a clear or unclear form. An integral indicator of the personnel ISC level can be used or providing audits of the organization's InfoSec system, and also in assessing the effectiveness of the implementation of methods for its development and improvement.

Further research may be conducted in the direction of simulation of a complex system for assessing the level of information security of the organization, taking into account personnel, technical, and technological components. Also important for the further implementation and development of the model is the creation of information technology for its support.

References

1. Heyden, L.: Informatsionnaya bezopasnost i Vseob'emlyuschiy Internet [Information Security and Comprehensive Internet]. Avtomatizatsiya I Sovremennyie Tehnologii (Electron. J.), 11, 29–31 (2013). https://www.cisco.com/c/ru_ua/about/press/2013/05292013g.html

2. Van Niekerk, J.F., Von Solms, R.: Information security culture: a management perspective. Comput. Secur. **29**(4), 476–486 (2010). https://doi.org/10.1016/j.cose.2009.10.005
3. Ngo, L., Zhou, W., Warren, M.: Understanding transition towards information security culture change. In: Proceedings of 3rd Australian Information Security Management Conference, pp. 67–73 (2005)
4. Furnell, S., Thomson, K.L.: From culture to disobedience: recognising the varying user acceptance of IT security. Comput. Fraud Secur. **2009**(2), 5–10 (2009). https://doi.org/10.1016/s1361-3723(09)70019-3
5. Da Veiga, A., Eloff, J.H.P.: A framework and assessment instrument for information security culture. Comput. Secur. **29**(2), 196–207 (2010). https://doi.org/10.1016/j.cose.2009.09.002
6. Flores, W., Antonsen, E., Ekstedt, M.: Information security knowledge sharing in organizations: investigating the effect of behavioral information security governance and national culture. Comput. Secur. **43**, 90–110 (2014). https://doi.org/10.1016/j.cose.2014.03.004
7. Schlienger, T., Teufel, S.: Information security culture: from analysis to change. S. Afr. Comput. J. **31**, 46–52 (2003)
8. Alhogail, A., Mirza, A.: A framework of information security culture change. J. Theor. Appl. Inf. Technol. **64**, 540–549 (2014)
9. Sherif, E., Furnell, S., Clarke, N.: An identification of variables influencing the establishment of information security culture, vol. 9190 (2015). https://doi.org/10.1007/978-3-319-20376-8_39
10. Bellman, R.E., Zadeh, L.A.: Decision-making in a fuzzy environment. Manage. Sci. **17**(4), 141–164 (1970). https://doi.org/10.1287/mnsc.17.4.b141
11. Saaty, T.L.: Relative measurement and its generalization in decision making: why pairwise comparisons are central in mathematics for the measurement of intangible factors - the analytic hierarchy/network process (PDF). RACSAM (Rev. R. Spanish Acad. Sci. Series A Math.) **102**(2), 251–318 (2008)
12. Fishburn, P.C.: Utility theory. Manage. Sci. **14**(5), 335–378 (1968). Theory Series

Author Index

© Springer Nature Switzerland AG 2020
A. Palagin et al. (Eds.): MODS 2019, AISC 1019, pp. 259–260, 2020.
https://doi.org/10.1007/978-3-030-25741-5

Printed in the United States
By Bookmasters